AF322658

THE ROADMAP TO ARCHITECT AND OPERATE
AI-DRIVEN MARKETING AT SCALE

THE MODERN CMO

HOW **MARKETING ENGINEERING** REDEFINES LEADERSHIP IN THE **AI ERA**

ROBERT PAINTER

BAUMANN | CHRISMER
PUBLISHING

Copyright © 2026 by Robert Painter

All rights reserved. No portion of this book may be reproduced, stored in a retrieval system, or transmitted in any form or by any means including electronic, mechanical, photocopy, recording, scanning, or other except for brief quotations in reviews or articles, without express permission. Inquiries may be directed to Baumann Chrismer, permissions@baumannchrismer.com.

This book is written as a source of information only. The information contained in this book should by no means be considered a substitute for the advice, decisions, or judgment of the reader's professional advisors. All efforts have been made to ensure the accuracy of the information contained in this book as of the date published. The author and the publisher expressly disclaim responsibility for any adverse effects arising from the use or application of the information contained herein.

The Baumann Chrismer Speakers Bureau can bring authors to your events and podcasts. For more information or to book an appearance, email us at business@baumannchrismer.com.

Cover Design by Bridget Painter

Hardcover: 979-8-234-01529-7

Softcover: 979-8-234-01530-3

Ebook: 979-8-234-01531-0

Library of Congress Control Number: 2026906645

First Edition

This book is dedicated to Gammie & Row, my loving parents who raised five children, who then had 12 children of their own and 4 grandchildren. Our extended family is now 32 and growing every year. And speaking about compounding growth, let's talk about modern marketing in the AI era…

CONTENTS

Preface

I've always been fascinated by how marketing shapes perception and guides how consumers make decisions. A brand that inspires, a value proposition that motivates, a campaign that resonates. As marketers, we live at the intersection of business priorities, customer sentiment, and cultural trends. And our goal is to earn a distinctive place in customers' minds while keeping our brands relevant.

From the outside, marketing feels intuitive and easy to do. When it works well, it feels right and almost obvious. From the inside, however, marketing is far more complex. There's incomplete information, strategic ambiguity, constant decision-making, complex collaboration, tight deadlines, countless reviews, and last-minute reworks. Marketing is always adapting and adjusting to get it done right.

I've spent my career in marketing, working across agencies and brands. I started in New York agencies, advising on positioning, branding, and creative for clients ranging from Diet Coke to product launches at IBM and Johnson & Johnson. What I remember most is the journey it took to make it happen. Along the way, there were moments of intensity and humility that only marketing delivers. I remember racing to deliver a final commercial to a network control room minutes before it aired during a live show. They were not happy. I remember being pulled over in Los Angeles while rushing to a Johnson's Baby shoot and explaining to the police officer that we were there "shooting babies." I never heard my creative director laugh harder. I remember traveling around the country to present NFL campaigns to local Coca-Cola bottler management teams. It felt like everyone had a different tactic they wanted to add to the program, and we delivered it.

Those early years taught me the power of positioning, creative excellence, and most of all, effective execution. They also reinforced that great ideas don't come easy. Marketing doesn't just happen. It's the result of sustained collaboration across teams and partners to bring ideas to life.

Those experiences deepened when I joined the corporate marketing team at IBM. The mission was to restore and modernize the brand. We worked closely with business units and regional teams to align and unify strategy and messaging. This included positioning IBM around the idea that the Internet was not just for browsing, but for real business, resulting in the launch of e-business. And we raised awareness of IBM's rapidly growing services business. This included the integration of IBM Global Services and the acquisition of PwC Consulting, introducing a new IBM capability focused on delivering business transformation onsite, offshore, or on demand.

I later brought these experiences to Cognizant, building a global corporate marketing function and helping to elevate the brand into the top tier of global technology consulting firm and digital service provider. At the time, we scaled Cognizant's extraordinary growth culture to more than 300,000 employees, without formal organization charts. The intent was to flatten the organization and encourage associates to deeply understand how the business operated and how to personally deliver exceptional client service.

My journey has spanned nearly every aspect of marketing, including research, strategy, planning, development, execution, and measurement. The focus has always centered on how marketing works to produce great outcomes. Marketers are naturally wired to collaborate, to figure things out, and to get the work done as a team. I'm deeply grateful for the colleagues, partners, and friendships built along the way.

That journey also placed me at the intersection of two worlds, marketing and technology. Over multiple eras, I witnessed how technology continually reshaped business and marketing. Now, with the arrival of Artificial Intelligence, we're poised for an entirely new level of productivity and scale.

AI has become the central topic in nearly every conversation I have with marketing and executive leaders. Early adoption has focused on enhancing existing marketing capabilities. At the same time, there is real tension across teams. How should AI be used? Who is responsible for our AI strategy? What are competitors doing? Alongside these internal questions, there's growing external pressures. Customers increasingly rely on it to guide their buying decisions, and they like it.

Along the way, it became clear to me that when marketing doesn't adapt to these shifts, it begins to lose what matters most, customer relevance. This is what pushed me to think differently about how marketing needs to operate. It became obvious that marketing was still being managed as a collection of activities, while the world around it had become increasingly intelligent and technology centric. We are asking marketers to compensate for an operating model that no longer fits the emerging AI environment.

That was the point where my career experiences converged. The creative and execution discipline I learned at agencies. The scale and strategy coordination required inside IBM. The speed and technology agility at Cognizant. All of it pointed to the same conclusion. Modern marketing performance comes from designing the structure that turns intelligence into action, repeatedly and reliably. That realization set the purpose of this book, to understand how AI is changing the way marketing works.

I'm deeply passionate about helping CMOs win in this new AI era. This isn't about chasing the latest technology or martech trends, but about helping leaders gain focus and confidence as marketing is fundamentally reimagined. In many ways, the future of marketing sits with CMOs leading today. They're deciding whether AI becomes another layer of complexity or the foundation of an intelligence operating model. That's why this book isn't about the latest models, platforms, or prompts. It's about the role of CMO leadership in the AI era.

The Modern CMO gives you a practical framework for architecting and operating marketing in the AI era. Above all, I hope this book leaves you with a sense of clarity. Clarity about this moment in marketing. Clarity about the leadership now required. And clarity about the opportunity to shape this new era of AI driven marketing. This is a management and leadership book for CMOs, CEOs, and senior executives who understand the magnitude of AI and want a clear path forward. Ultimately, it shows how the future of marketing is engineered.

I want to credit the people who helped make this book possible. I'm appreciative of the many marketing leaders who shared their stories and perspectives on AI. I'm grateful for colleagues who kept me motivated and validated the direction. I want to acknowledge Tanmay, for grounding the architecture in technical underpinnings. Natalie, for her invaluable edits and precision. Special thanks to Bridget for bringing the ideas to life visually and driving production with great expertise and care. And most of all, thank you to my wife Jill, who supported me endlessly while I disappeared into this work every day for many months.

Introduction

A CMO I've known for years pulled me aside after a work session and said, "I feel like I'm running an organization built for a world that doesn't exist anymore." She had the vision and confidence and understood exactly what was changing around her. But the expectations of AI were rising faster than her team could support. And she isn't alone. Many CMOs now feel that AI's acceleration is outpacing their teams' skills and capabilities.

Every era of business reaches a moment when the rules of advantage shift, and for marketing that moment is now. Artificial Intelligence is no longer in the background. It's become the environment we all work in, and you can feel its impact in the way customers use AI to explore, compare, and choose. And in the growing pressure on your team to keep up.

Marketing once ran on creativity, intuition, and repetition. These strengths still matter and they always will, but they cannot carry the full weight of modern expectations. CMOs who excel in this new arena will be the ones who understand that the real shift is about how marketing must run when AI becomes the fabric of every decision, every workflow, and every customer moment. The market is moving faster than most teams can respond, and leaders are now in an environment that feels both urgent and unfamiliar.

This isn't a book about technology. Keeping up with new AI developments already feels overwhelming. New LLM model (e.g., ChatGPT, Claude, Gemini, xAI) versions emerge every month, and compelling platforms often create more noise than value. The pace of technology innovation is definitely intense, and it's more than any leader can track with confidence.

This is a book about how CMOs make sense of the world around them and guide their teams through it. The book explains how marketing runs when the market demands more foresight, productivity, and speed than the traditional model can deliver. And it's about the leadership vision and the decisions that shape performance in this AI era.

In my conversations with CMOs across many industries, I hear the same questions, which aren't about prompts or tools. They're often questions about leadership and the future of the function. Leaders want to know what competitors are doing with AI, and how to strengthen the conversations they have with CEOs and Boards. They want better oversight of their data, and insight on how martech should align with intelligent workflows. They want to understand how AI driven search will affect brand visibility, and how to redesign roles and teams for this landscape that is shifting quickly.

If these questions feel familiar, this book is for you.

Marketing has grown in layers over the years. New strategies and channels appeared. New workflows, tools, roles, and metrics followed. Each addition brought capability and responsibility, but the operating model didn't evolve at the same pace. It was built for a slower and more predictable environment where planning cycles matched the speed of the market, customer journeys followed clearer paths, and repetition could build relevance and trust. That era is gone, and every CMO is feeling it.

Marketing isn't struggling because teams lack expertise. It's struggling because a legacy structure cannot support the demands of this new environment. The pace of AI innovation is too fast for isolated pilots to matter, and the noise surrounding new martech features pulls attention away from the deeper work.

The Modern CMO offers a roadmap for addressing these issues by shifting the focus from what marketing produces to how marketing operates. It introduces the AI Marketing Operating System (AIM OS) and the discipline of Marketing Engineering as a practical way to build a function capable of meeting today's demands.

This roadmap matters because modern marketing can no longer be sustained by big ideas or heroic team effort. It needs an operating model designed for an AI foundation. And it needs a new discipline that brings marketing and technology together. This book is for today's CMO who understand AI's acceleration isn't only a technology shift. It's an organizational shift.

This book is built as a progression. Each part prepares you for the next, creating a clear path from understanding today's shift to running a modern marketing model at scale.

We begin with **Part I: Marketing Collides With the AI Era.** This describes what's happening now and why the old model can't keep up. The section reframes the reality every CMO is living in. It shows how intelligence is reshaping customer behavior, why traditional marketing structures are slowing down, and where the operating model no longer matches the environment. By the end of Part I, you'll see clearly why marketing feels heavier and more complicated than it should and why incremental fixes no longer work.

In **Part II: Building the AI Marketing Operating System,** we dive into what the new model looks like and the discipline required to build it, introducing the architecture modern marketing needs. You'll learn how the AI Marketing Operating System works, why Marketing Engineering has become the missing discipline, and how CMOs must lead marketing as a system rather than a collection of activities. This is the roadmap delivering the AI driven structural clarity most organizations lack.

Then, **Part III: Running the AI Marketing Operating System** shows how the system works in motion and how it changes performance. This section makes the model very tangible. It shows how data and customer signals move through your system, how decisions accelerate, how activation happens in real time, and how learning compounds over time. It clarifies the metrics that matter, the role of data, and how efficiency becomes a platform for growth. This part turns the operating model from an architecture into something you can run every day.

Finally in **Part IV: Marketing Leadership in the AI Era,** we discuss how leaders create momentum and drive the organization forward. The final section explains why modernization often stalls and what leaders can do to prevent the system from sliding back into old behaviors. You'll learn how to break the momentum traps that slow teams down, how to reinforce new patterns, and how to lead with clarity in a world where intelligence drives expectations.

By the end of the book, you will see the CMO role differently.

You will understand why modern marketing cannot scale through effort, talent, or tools alone, and why operating structure has become the central responsibility of the CMO and their leadership team. You will leave equipped with the architecture and the clarity needed to build a marketing function designed for the AI era.

This work matters. It's time for a new marketing model. It's time for a discipline that engineers competitive advantage. And most importantly, it's time for CEOs and CMOs who understand that the real shift underway is structural.

Use this book to anchor your team in the right AI marketing mindset. Encourage your leadership teams to read. And use the reflection questions in each chapter to spark alignment and action.

If marketing feels more complex than it should, or if your team is working harder than the results show, and if your AI pilots aren't creating real capability, this book will help you move forward. It will give you the roadmap to build your AI Marketing Operating System and the discipline to run modern marketing with confidence.

The CMO is no longer defined by *what* marketing produces. The Modern CMO is defined by *how* marketing works in the AI era. And that's where this journey begins.

Let's get started.

PART I

Marketing Collides
With The AI Era

1

How Intelligence Rewrites the Playbook

Competing in an AI defined marketing environment

If you're a CMO today, you can probably feel it. The scope of marketing keeps expanding. Teams are busy. Data is growing. Promise of new technologies appear every day. And yet, it doesn't always feel like marketing is fully in control of its outcomes the way it used to be. Consumer decisions seem to take shape earlier. Influence feels harder to hold. It seems the gap between effort and impact has widened in ways that are difficult to explain.

Most CMOs I know are doing the right things. They're building strong teams, investing thoughtfully, and leaning into change. What's evolving isn't the quality of leadership or the importance of marketing. It's the landscape marketing must navigate today. The environment has shifted quietly, even as day-to-day responsibilities still look familiar. And this gap is widening.

Customers now move through experiences that are quietly adapting. Every interaction feeds a larger system. Purchases refine recommendations. Behavior reshapes offers. And journeys adjust in real time. When these experiences are seamless, technology fades into the background. They feel personal and effortless.

This shift is already visible in search. As AI generated summaries and overviews become common, customers are responding quickly. Search feels faster. Answers feel more relevant. Decisions feel easier. For customers, this efficiency is welcome. For marketing leaders, it signals a deeper change in how brands are chosen.

Visibility no longer depends on winning attention or driving a click. It depends on whether intelligent systems can understand, interpret, and trust what an organization offers. Brand relevance is no longer shaped only by campaigns or content. It's increasingly granted, or withheld, by the models generating the answer.

The implications extend beyond where marketing appears. Customers gain leverage while forming new dependencies. They spend less effort searching yet rely more heavily on the AI systems guiding their decisions. Trust shifts away from individual brand claims toward the reliability of the AI results. Loyalty becomes less about habit or promises and more about whether those systems consistently deliver clarity.

What matters now is not only how compelling your positioning sounds to customers, but also how clearly competitive benefits can be interpreted by machines. Most organizations can feel this shift, even if they struggle to adapt.

One retail bank saw this firsthand as customers began using AI powered search to evaluate services before visiting branches. Systems summarized reviews, compared rates, and generated shortlists. While the company's messaging resonated well with customers, it was largely invisible to the models shaping early decisions.

Before customers visit a website, speak with sales, or engage directly with a tactic, many decisions are already taking shape. Increasingly, those decisions form inside AI generated results that summarize options, compare capabilities, and narrow the choice on the customer's behalf. In those moments, the first evaluation of a brand is often made by a system rather than a person.

This pattern is now common across industries. Intelligence operates on both sides of the interaction. Customers use it to decide. Marketers use it to perform. This shift has changed where marketing happens and how performance is determined. Advantage no longer comes from just better campaigns or larger budgets. It comes from the performance of the intelligent operating models beneath them.

This is the gap *The Modern CMO* addresses.

Marketing now operates inside an AI driven environment where performance depends on how demand is sensed, how priorities are set, how execution is activated, and how learning compounds over time. Results come less from individual outputs and more from the structure that allows intelligence to translate into precise action.

Relevance is no longer something you claim through messaging. It is something the system produces through performance. Marketing is being rewired around intelligence rather than tools, around operating models rather than outputs, and around systems that can perform continuously as conditions change. This is the shift the modern CMO must confront.

Relevance Rewired

Marketing has always carried a simple promise, to close the gap between what companies offer and what people want. At its best, it is a discipline of relevance guided by instinct, sharpened by analysis, and brought to life through creativity. When those forces align, selling feels effortless, buying feels easy, and brands earn their position in the customer's mind. That core promise hasn't disappeared, but the foundation beneath it has changed.

Instinct, analysis, and creativity still matter, but they can't keep up on their own. Instinct can't interpret millions of signals firing across digital channels every second. Analytics can't keep pace with behaviors that move in real time. Even the boldest creative idea falls flat when buying paths are reshaped mid-journey. The levers that once drove demand and growth weren't built for the velocity and complexity of the AI era. But treat intelligence as the operating fabric, and relevance becomes more resilient.

Too often, however, AI is framed as a feature on a platform. It's seen as a faster way to produce content or a path to productivity gains. These uses miss the point. Intelligence isn't an add-on. It's a capability. It's the connective circuitry that must now run through the entire marketing discipline. That's why the real question is no longer whether AI can generate content, personalize messages, or automate tasks. The real question is whether marketing can be reimagined around the deeper, continuous layers of intelligence.

Intelligence isn't a tool. It's a capability.

Seen through this lens, the structure of traditional marketing breaks down. You can no longer launch messages and hope they land. Algorithms filter them. AI agents interpret them. By the time a value proposition appears, the customer is already several steps ahead. This shift rewires the customer relationship, reshapes how teams operate, and changes how marketing fundamentally works. What once looked like plans, campaigns, and reports becomes a dynamic model that senses, decides, and acts alongside customers.

This isn't incremental change. It's a step into a new era. Fueled by intelligence, marketing becomes a different way to build relevance, earn preference, and sustain growth. That is both the opportunity, and the mandate, of the Modern CMO.

The Five Superpowers of Machine Intelligence

When most people hear "AI," they think of faster content creation, automated workflows, and incremental cost savings. Those associations are remnants of the first wave of hype. They frame intelligence as efficiency and productivity promising the same work faster and cheaper.

The shift underway is far more consequential. It's not about efficiency. It's about how new dimensions of intelligence are being applied across decision-making, coordination, and execution. These capabilities change how marketing operates, how customers choose, and how advantage is built and sustained.

Seen clearly, machine intelligence introduces a set of very distinct strengths for marketing. We call them the five marketing superpowers. They are five new foundational capabilities that were previously impossible to achieve at scale. They aren't platforms or software features. They're operating advantages. And together, they explain why marketing is moving from effort-driven execution to intelligence-driven performance.

The five marketing superpowers of machine intelligence:

1. **Sensing and Learning:** The first power is the ability to absorb signals, detect patterns, and improve continuously as new data flows in. Streaming services demonstrate this every day, translating millions of listening behaviors into playlists that feel personal without requiring explicit instruction. In marketing, sensing and learning moves beyond static surveys or periodic studies. Systems learn continuously from first-party data, customer interactions, and unstructured signals to understand behavior as it changes in real time.

2. **Reasoning:** The second power is the ability to interpret context, weigh trade-offs, connect signals across complexity, and recommend direction. Logistics platforms use this to calculate delivery routes by balancing time, traffic, and cost in milliseconds. In marketing, reasoning enables clearer choices like which audience to engage now, which message to prioritize, and which markets deserve focus. It allows intelligence to guide decisions instead of forcing teams to rely on assumptions or benchmarks.

3. **Decisioning:** The third power is the ability to select the next best action instantly at scale. E-commerce platforms demonstrate this through dynamic pricing that adjusts in real time as conditions change. In marketing, decisioning replaces fixed plans with responsive allocation. Budgets shift as performance unfolds. Offers adapt to context. Actions respond to signals rather than following journey map steps or waiting for management reviews to catch up.

4. **Autonomous Action:** This power is the ability to act continuously without manual intervention while optimizing performance in motion. Programmatic advertising already executes millions of bids per second. That pattern is expanding. Journey content adjusts as customers respond. Loyalty incentives evolve with behavior. Execution no longer pauses for approval. The system acts autonomously, removing repetitive work.

5. **Generative Creation:** And the fifth power is the ability to produce new content, experiences, and solutions tailored to each audience and moment. In design and retail, AI already generates product concepts and visuals that adapt to style and preference. In marketing, generative creation means content that evolves with context. Messages reshape themselves based on audience, channel, and timing, allowing relevance to be maintained without manual rework.

Each superpower delivers significant value on its own. But together, they form an intelligence fabric that runs through the entire marketing discipline. This is the path ahead. Insight becomes continuous. Action becomes responsive. Experience becomes adaptive. And the system learns as it operates, compounding performance over time. For CMOs, these capabilities are no longer aspirational. They're real. And they redefine how marketing works.

How Intelligence Reframes Marketing Capabilities

The simplest way to understand this shift is to ask a single question: does deeper intelligence make this better? In life sciences, recruiting, product development, logistics, and financial management, the answer is consistently yes. Systems that can analyze, reason, decide, and act with greater intelligence become more effective, more resilient, and more central to how work gets done. Marketing is no exception.

Consider the core capabilities. Would customer insight improve if it drew from many more active signals and real time context rather than static, historical data? Yes. Would content perform better if it adapted continuously to buying behavior instead of following fixed calendars? Yes. Would media and attribution improve if budgets were allocated dynamically toward actual performance rather than pre-set plans? Again, yes.

The same pattern repeats across the discipline. Digital experiences become responsive rather than scripted. Account-based marketing shifts from targeting lists to sensing account readiness. Customer journeys stop behaving like linear maps and start operating as adaptive pathways. Media investment moves from quarterly adjustments to continuous allocation. Positioning evolves from an annual exercise into an ongoing dialogue with a market in motion.

And as intelligence permeates these outward-facing capabilities, it reshapes the internal mechanics of marketing as well. Budget and resource management become adaptive rather than negotiated once a year. Operating models and organizational design respond more fluidly to changing conditions. Leadership enters the boardroom with system-level clarity instead of backward-looking reports. Even brand reputation becomes more governable when intelligence can detect weak signals before they escalate into visible risk.

Artificial intelligence doesn't replace marketing. It elevates it.

Every capability grows stronger when viewed through the lens of intelligence. This isn't because marketing becomes more automated, but because it becomes more precise, more responsive, and more aligned to how customers now behave. Intelligence doesn't sit beside the function. It permeates the structure, changing how relevance is created and performance is sustained

Exhibit 1: **Capabilities Through the Lens of Intelligence**

Capability	How Intelligence Elevates Marketing
Content Marketing	Transforms content from static assets into adaptive experiences that refine in real time.
Program Management	Moves to dynamic market mix modeling that autonomously updates based on performance.
Demand Generation	Agentic systems coordinate cadence and creative tests autonomously across channels.
Media Strategy	Treats media like real time capital allocation using privacy-forward signals.
Product Marketing	Uses market simulators to pressure-test positioning and price before launch.
Brand Reputation	Early-warning models monitor customer signals to identify issues before they escalate.
Martech Strategy	Shifts from platform stacks to an integrated, intelligence-based operating fabric.
Websites	Edge inference makes experiences intent-adaptive and instantly personalized.
Social Media	Reads audience sentiment, automates content creation, conversation agents engage.
Search	Shifts from keyword lists marketers optimize to influencing AI generated overviews.

The Shift from Mix to Machine

It's one thing to make individual marketing capabilities more intelligent with insights that predict, content that adapts, and journeys that respond in real time. But the true shift is how these intelligent capabilities create even more expertise and leverage when connected as a system, turning fragmented efforts into a growth engine. This level of marketing performance has been promised in theory for years but rarely delivered in practice.

Other disciplines illustrate this transformative opportunity. A factory isn't just machines; it's the system that synchronizes them into continuous flow. A financial market isn't just individual trades; it's the system that links signals, orders, and liquidity into dynamic equilibrium. A supply chain isn't just vehicles and warehouses; it's the system that coordinates production, transport, and demand into resilience. In every case, intelligence in one part of the system improves performance. But intelligence across the system delivers compounding impact.

Intelligence only scales when leadership gives it purpose and guardrails.

Marketing is no different. Today, many teams experiment with artificial intelligence in silos like predictive models in customer analytics, generative tools in creative production, or recommendation engines on web sites. Each delivers value in isolation, but these actions are merely augmenting and optimizing the current traditional marketing structure and operating model. For example, insights still sit idle in spreadsheets instead of triggering action. Journeys stall at organizational handoffs. Opportunities are scored with passive signals, and budgets sit underutilized for months across team departments.

Consider a global retailer. They invested heavily in AI platforms for creative production and customer service chatbots. The technology worked as campaigns were produced faster, service queries resolved more efficiently. But sales, however, didn't move. Targeting was static, media spend was anchored in quarterly plans, and customer journeys broke at point-of-sale. ROI increased when they fused intelligence across dynamic targeting that fed content systems, media spend that reallocated daily, and incentive journeys that adapted in real time. The breakthrough wasn't a new tool. It was an orchestrated system.

Intelligence in one capability improves marketing.
Intelligence across the system completely transforms marketing.

When capabilities and workflows remain fragmented, marketing works harder, but not smarter. When they are integrated in an intelligent system, the rules change. Signals flow seamlessly. Models reason. Decisions trigger actions. Outcomes feed back into the system. Each interaction improves the next. The promise of marketing through the lens of intelligence isn't smarter parts, but a smarter whole. The customer experience is still shaped by marketing. But beneath it runs an intelligent system designed to sense, decide, allocate, and act at the speed of AI.

The Scale of the Shift

The world is funding machine intelligence at an unprecedented scale. Global investment in AI is already measured in the trillions, and it is accelerating faster than almost any major technology transition in modern history.

Enterprise spending alone is projected to reach roughly $2.5 trillion by the end of 2026 (Gartner), driven by software, services, and infrastructure adoption. Forecasts consistently show that growth continuing well beyond 2026 as intelligence becomes embedded into core operations rather than deployed as isolated tools.

This growth isn't speculative. Large enterprises and hyperscalers are expected to invest over $3 trillion in AI infrastructure through 2030 (Citibank), reflecting the computing scale, data, and model capacity required to deploy intelligence at enterprise level. The capital is flowing into platforms, cloud infrastructure, AI services, and operational integration.

Private investment reinforces the same trajectory. Annual corporate and private AI investment has already exceeded $250 billion (Stanford AI Index), with enterprise adoption now outpacing early-stage experimentation. Government spending, while smaller in absolute terms, is also accelerating as nations race to fund national AI strategies, data infrastructure, and sovereign capability.

What makes this surge different isn't just its size, but its compression in time. The dot-com boom reshaped markets, but at a fraction of today's scale. Mobile and cloud unfolded over more than a decade. AI, by contrast, is expanding simultaneously across startups, global enterprises, and governments. This is compressing what would normally be a ten-year transition into just a few years.

By the second half of this decade, annual AI spending is expected to rival the scale of the world's largest economic commitments. Global defense spending totals just over $2 trillion annually (SIPR). The global advertising industry approaches $1 trillion (GroupM). AI is on a trajectory to stand alongside these sectors as a foundational layer of the global economy.

The implication is that intelligence is becoming infrastructure. And when intelligence becomes marketing infrastructure, performance is no longer determined by tools or tactics. It's determined by the operating models built to harness it.

Artificial intelligence represents the largest capital allocation shift in the modern economy.

That scale matters a lot because it reframes the role of marketing leadership. For decades, technology spend was treated as an IT concern, while marketing budgets focused on programs and media. When trillions of dollars flow into intelligence, that dynamic changes quickly.

For CEOs, the implication is unavoidable. Capital at this scale produces disruption at speed. Competitive advantage will move quickly to those who operationalize intelligence effectively. For CMOs, the mandate is certain, intelligence will redefine marketing, and capital expense will follow operating models that turn it into performance. For marketing professionals, this is a major opportunity. Intelligence doesn't replace your work. It amplifies it.

While the precise investment totals will vary, the conclusion does not. Machine learning and AI are no longer experiments. They represent the single largest reallocation of capital in the modern economy. And how organizations apply that intelligence will determine who leads and who falls behind.

The Modern Consumer Implication

The surge of investment into artificial intelligence isn't just reshaping industries. It's reshaping how customers experience the world. Consumers still engage with brands, but increasingly they do so inside AI mediated results that anticipate needs, interpret intent, and adapt in motion. This shift carries consequences far beyond personalization.

Customers now move through journeys that are always on and highly responsive. Every interaction becomes part of an ongoing exchange. Purchases refine recommendations. Behavior reshapes offers. This process continually adjusts in real time. When designed well, these experiences don't feel like technology at all. They feel very personal and highly relevant.

The change is already visible in search. As AI generated summaries and overviews become more common, users are responding positively. Search feels faster. Answers feel more contextual. Decisions feel easier. For customers, this efficiency is a welcome improvement. For marketers, it introduces a more uncomfortable reality. As people rely on AI to synthesize choices, they encounter fewer open result pages and fewer visible brand options. Relevance is no longer earned by ranking higher on a list of links. It's increasingly granted or withheld by the LLMs shaping the answers.

In the AI era, visibility is no longer what you broadcast. It is how intelligence interprets and represents your brand in context.

These implications are profound. Customers gain leverage, but they also form new dependencies. They're freed from the effort of searching, yet more tightly guided by the systems helping them decide. Trust begins to shift away from individual brand claims toward the reliability of AI guidance. Loyalty is no longer driven by habit alone. It depends on whether AI experiences consistently deliver clarity.

With trillions of dollars flowing into artificial intelligence each year, consumers will move through LLM-powered engines as part of everyday life. Every interaction will carry an expectation of relevance, transparency, and care. The real test of intelligence won't be what it automates, but how it shapes the experience customers come to rely on. That shift will redefine what relevance means and who gets to earn it.

Early Signs You're Missing the Shift

The evidence is almost invisible at first, and then impossible to ignore. Campaigns that once surged out of the gate begin to stall, even when the budgets behind them remain intact. The message hasn't grown weaker, but now the place where customers make decisions are moving faster than the cycles used to plan them.

Inside the organization, the warning signs accumulate. Customer journeys feel fractured. Personalization remains shallow, as signals slip between disconnected martech platforms. Loyalty erodes quietly, not because customers dislike the brand, but because their experiences don't align with their expectations as repeat customers.

And then there is the boardroom. Questions that once centered on awareness or opportunities now focus on return. The numbers presented sound like activity versus outcomes. The answers speak to effort rather than evidence. The board senses the gap in how artificial intelligence is failing to reimage marketing.

These aren't failures in the traditional sense. They are the early signals of a marketing function still operating as if campaigns, silos, and instincts can keep pace with AI that learns and adapts in real time. They are the signs of treating artificial intelligence as an experiment rather than the fabric of operations. For CMOs, the warning is clear. Marketing won't be replaced, but it can be sidelined if it doesn't claim its role as the architect of the system that shapes the environments where growth is now decided.

The Modern CMO Implication

If consumers are exploring and using AI, then your leadership approach cannot stay the same. Marketing has always carried one central promise to make a business relevant to the customers it serves. You've delivered on that promise through creativity, instinct, and analysis. Those strengths are still on your side, but AI expands what they can do. Creativity becomes experiences that adapt in real time. Instinct becomes models that learn continuously. Analysis becomes customer intelligence that compounds across the enterprise. The shift is irreversible. Your work must evolve from a sequence of activities into a system of intelligence that is designed to anticipate, adapt, and earn trust at scale.

Think of the billions flowing into artificial intelligence as an invitation to the future. And this is addressed to you. No other executive sits closer to the customer, closer to market signals and closer to cultural shifts. No other executive is better positioned to translate intelligence into relevance, to turn data into action, and to make your company visible in new places where decisions are already being shaped.

History offers some lessons from past technology shifts. Lessons like the rise of web, the digital wave, and the spread of cloud. Each completely reshaped industry. But none carried the sheer impact of what AI represents today. This is more than a side project. It'll be the operating fabric of the economy.

That's why your role is being elevated. Marketing becomes the system of relevance, showing the way intelligence translates into trust with customers and confidence with leaders. The future won't ask whether you ran enough campaigns. It will ask whether you built the system that made relevance inevitable. The opportunity couldn't be more consequential. So, the question is: *Are you ready to be the modern CMO your business now needs?*

Summary

Global capital is moving decisively toward artificial intelligence. Trillions of dollars are now being committed to intelligence as infrastructure, not experimentation. Markets are signaling clearly that performance in the next decade will be determined by how effectively organizations operationalize intelligence at scale.

This shift doesn't reward one more tool, platform, or pilot. It rewards leaders who redesign how work gets done. In AI driven environments, advantage comes from operating models that can sense demand, interpret context, decide priorities, allocate resources, activate execution, and learn continuously. When intelligence is embedded into the way decisions are made, relevance is sustainable and performance becomes repeatable.

Marketing sits at the center of this shift because marketing is closest to the customers where trust and choice are formed. The risk isn't that marketing becomes less important. The risk is that it's bypassed by intelligent capabilities that already understand customer intent before marketing ever engages.

Leadership Review

1. Are we treating AI as a tactical accelerator, or as a structural redesign of how marketing creates performance?

2. What is our plan to move from pilots to an operating fabric that improves every cycle?

3. How are we ensuring visibility in AI environments and answer engines where customer choices are increasingly made?

4. What evidence shows that intelligence is improving customer relevance, not just execution efficiency?

5. Do we have clear ownership, guardrails, and escalation paths to ensure AI strengthens trust as it scales?

2

Traditional Marketing Can't Keep Up

*Why yesterday's marketing approaches
are holding you back*

"I think we need a new CMO," said the CEO of an insurance company I was meeting with. Here we go again, I thought. "What makes you think that?" I asked.

"Our customers are shopping differently now," she said. "They don't come to learn anymore. They've already asked AI which companies have the best policies, the lowest fees, and the strongest ratings. By the time they reach our site or a branch, they're already ahead of our sales agents."

AI powered assistants now help consumers summarize coverage, compare options, and generate shortlists in seconds. While the company's comparison tools and messaging resonated well with customers, they were largely invisible to the large language models shaping early decisions.

Her frustration was understandable. I asked, "How are you adapting to this?" She paused. "We're not, really. Marketing knows it's happening, but we don't know how to respond yet. We don't even know how these machine decisions are being made."

I explained that marketing had been designed to influence people, not to be interpreted by intelligent systems. The execution model behind the brand simply wasn't legible to the systems now shaping decisions. Judgment was being formed by intelligence that never attended their webinars, never scrolled their landing pages, and never clicked an ad. A content audit later confirmed these issues. Coverage details appeared inconsistently across sources, making them difficult for AI systems to interpret accurately.

I pressed further. "So what is marketing doing that's working?" She replied, "The team runs many good projects. But it feels like we're always reacting. We're chasing issues as they surface, especially with these new AI answer engines. Customers are comparing us in seconds, and our branches feel like they're falling behind."

That's when the conversation shifted from leadership to structure. We talked about how marketing could anticipate customer decisions instead of reacting to new behaviors. "It's not your CMO," I said. "Your marketing model is outdated. It wasn't built for this era of intelligence. But it can be." She leaned in. "How do we do that?"

That was the moment the discussion stopped being about replacing leadership and started being about reimagining *how* marketing works. It's also the question every CMO and CEO should be asking now.

How does modern marketing run?

Because once you view marketing through the lens of the AI era, the cracks in traditional models become visible. You start to see the silos and the inefficiencies baked into everyday activities. Most teams aren't underperforming because they lack talent. They're underperforming because the structures they operate within were designed for another time.

Being a CMO today is fundamentally different from what it was even a few years ago. The role has expanded in scope and complexity. The lines between marketing, data, AI, technology, and operations have blurred. Expectations continue to rise, while the path to meeting them feels less defined.

What's changed the most is velocity. Buying behavior shifts faster. Economic conditions remain volatile. New competitors emerge regularly. New technologies multiply. And leadership expectations accelerate. But most marketing structures aren't built for this kind of pace. They were built for linear campaigns, rigid planning cycles, and slow handoffs. So, while AI promises innovation, it often meets many roadblocks before it delivers acceleration.

What comes next isn't incremental improvement. It's a new operating model designed to move at the speed of intelligence.

It's difficult to predict exactly what marketing will look like in the years ahead. The pace of change is compounding. New capabilities emerge before teams have absorbed the last wave. This creates a growing gap between what's possible and what's practical, forcing marketing leaders to rethink how the function operates.

The question is no longer whether technology will transform marketing. It already has. The real question is whether marketing will transform itself fast enough to keep up.

This book is written as a guide and a bridge for both CMOs and CEOs. It's for CMOs carrying the weight of marketing's complexity and looking for a more sustainable way forward. And it's for CEOs seeking greater clarity and return from marketing investments, even as expectations rise. The goal is to create shared understanding between both roles, because the future of marketing performance clearly depends on a tighter, more strategic partnership.

Imagine headlines like *"With AI, CMOs Become Architects of Growth"* or *"Marketing Becomes the Organization's Intelligence Advantage."* These are the horizons this book brings into focus. The future of marketing leadership will be intelligent, adaptive, and continuously tuned to customer behavior. Getting there requires more than ambition. It requires rethinking how marketing operates, and not just what it produces. Because in the AI era, the systems behind the scenes matter as much as the go-to-market strategies.

Wired to Adapt

To understand where marketing is headed, we need to take an honest look at where it stands today. Over time, layers of innovation, processes, and roles have been added in the name of progress. Instead of clarity and scale, many teams now carry significant cost and complexity. On some days, it can feel like a miracle that anything gets produced and launched at all.

Success today depends as much on orchestration as it does on insight and creativity. Marketing hasn't lost its magic. But making that magic work now requires more coordination across teams, tighter calibration across tools, and constant rerouting of workflows under pressure. The work gets done, but it takes more energy and resources than it should.

Adaptability has always been one of marketing's defining strengths. Marketers don't resist change. We run toward it. Our work lives at the intersection of culture, technology, and human behavior. We're expected to see around corners, respond in real time, and translate noise into insight and action before the rest of the organization even knows something has shifted. We do this to keep brands relevant, products desirable, and experiences aligned with what customers want.

When markets change, customer needs evolve, or new sentiments emerge, marketing is expected to move first. While other corporate functions rely on age-old playbooks, marketing is wired for agility. We respond to new customer preferences, emerging trends, and evolving priorities. We adapt and move forward. Strategies are built in real time, often without perfect data.

This adaptability has always been one of marketing's greatest strengths. What holds us back isn't the pace of change, it's when the systems, tools, or internal structures around us can't keep up. It's when our teams are stretched thin across too many priorities, that agility can become a strain.

Even the strongest strategies stall when structure fails.

This tension between marketing's natural adaptability and its structural constraints has defined much of the past two decades. And yet despite the challenges, marketing has continued to find a way forward. Reinvention is part of marketing's DNA, and the function has evolved alongside every major wave of innovation, creating new channels, adopting new tools, and experimenting with new ways to engage customers.

But each wave left something behind. New strategies, roles, and approaches were layered on top of the last. Over time, those layers accumulated. As marketing expanded to keep pace with innovation, it became more fragmented and a lot harder to orchestrate.

Today, marketing stands on the edge of another major shift, powered by AI and intelligent systems. The opportunity is significant. But realizing it will require more than merely adapting. It requires rethinking how marketing operates.

Marketing's Accumulated Eras

To understand how marketing reached this point, think of this section as a structural history. Each time a new technology or management idea arrived, we added it on top of what was already there. This included a fresh mindset, new skills, and different rhythms for how the work should run. Over time, those layers became today's norms. They gave us more capability, but they also added more cost and complexity.

What follows is a quick walk through of the layers, not for nostalgia's sake, but to understand why a redesign is needed for the AI era.

The Mass Media Era. This was the age of television and print, when marketing told stories from the top down, built brand equity in living rooms, and shaped culture through sheer reach. It was the golden age of big ideas and big media buys, where campaigns were linear and tightly managed. The strategy centered on positioning and unique selling propositions. Marketing was about airtime, audience ratings, and crafting the perfect 30-second spot. And for a long time, it worked.

It created brand icons. It created new categories. It was built for a world where attention was centralized, and media was scarce.

This era shaped how marketing operated, and who it valued. Creative directors, copywriters, and media buyers held the most influence by crafting creative platforms and negotiating prime airtime. Roles like Brand Manager, Advertising Executive, Strategic Planners, and Media Planner, dominated agency and brand teams. Success hinged on creative instinct, brand management, and relationships with major networks and publishers. It was a time when marketing was more art than science and was driven by ideas and powered by channels that could shape public perception at scale.

The IMC Era (Integrated Marketing Communications). Coined by Don Schultz, this wave marked a strategic shift from siloed ideas to coordinated storytelling. The personal computer (PC) was a foundational enabler. It gave marketers direct access to tools like spreadsheets for audience segmentation, campaign planning, and early data analysis, laying the groundwork for customer relationship marketing (CRM). As customer data moved from filing cabinets to desktop databases, marketers began building more targeted, coordinated campaigns across direct mail, promotions, and point-of-sale, accelerating the shift towards integrated, relationship-driven marketing.

For the first time, marketing leaders began intentionally aligning advertising, PR, direct mail, and promotions into a unified brand journey. The goal was consistency across every touchpoint, ensuring what customers saw on TV matched what they read in print, experienced in sales collateral, or encountered at point-of-sale. Integration became a competitive advantage and brands orchestrated campaigns across channels. It was a meaningful evolution. Marketing became more cross-functional, more connected, but it was still largely governed by linear plans, calendar-driven campaigns, and long lead times. The structure improved, but the system remained sequential.

This new level of coordination also shifted both talent and team structure. Traditional roles like brand manager and media planner were joined by emerging titles such as CRM Strategist, Direct Marketing Analyst, Customer Insights Manager, and the essential Integrated Marketing Communications Executive. Org charts now reflected the need for cross-channel coordination and data driven execution. Integration required talent that could think across functions, drive integrated planning, and turn umbrella brand frameworks into cohesive customer experiences.

The Internet Era. The web hit fast, and it seemed marketing was reborn online overnight. Teams built global websites. Search became the new storefront. And digital advertising, banner ads, and pop-ups promised precise targeting, quantitative tracking, and immediate feedback.

A new vocabulary emerged like click-throughs, impressions, bounce rates, conversion funnels. Google started to matter as much as shelf space at retail. Digital advertising brought precise targeting, real time tracking, and measurable performance. It was fast, direct, and data driven. And it marked the beginning of marketing's online transformation. We weren't just reaching audiences, we were watching them respond. And that sense of control and accountability reshaped the expectations of every marketing leader. Suddenly, the customer funnel went online. And what had once been a one-way broadcast to the masses became an interactive, data-fueled experience. Brands now had digital front doors open 24/7, with customers arriving from all directions at all hours.

This web explosion reshaped strategies and entire org charts. Traditional roles gave way to a new generation of specialists fluent in code, content, and conversion. Roles like Web Producer, SEO Manager, Display Media Buyer, and Email Marketing began to populate marketing departments. Analytics and performance tracking became core competencies, giving rise to roles like Conversion Rate Specialist and Marketing Analyst. Designers learned UX. Writers became content strategists. Marketers were suddenly expected to understand web traffic, tagging, and attribution models. It was a completely new mindset that was experimental and metrics driven.

The Digital Era. Another inflection point. Social platforms exploded. Mobile became the default screen. Content shifted from supporting material to the main event. Brands were publishing, tweeting, posting, streaming, and producing more content in a week than they once did in a year. The pace was relentless, and the noise was constant. Calendar driven work became a living ecosystem where campaigns evolved in real time and relevance was measured by engagement.

Programmatic advertising reshaped media buying, replacing manual negotiations with algorithms and collapsing planning cycles into milliseconds. Customer journeys fractured. Attention spans shrank. And the job of marketing became less about delivering a message and more about sustaining customer engagement by constantly adjusting to likes and cultural shifts. In this era, success was driven by many interactions that needed to land with precision.

As channels multiplied and content became the currency of engagement, new roles emerged to meet the specificity of the moment. For example, titles like Social Media Manager, Content Strategist, Community Manager, and Programmatic Trader became essential to teams. Video producers joined copywriters. Influencer leads sat next to media buyers. Real time analytics required roles like Engagement Analyst and Digital Insights Lead to track sentiment, optimize content, and respond to trends as they happened. Marketing was no longer just creative and media, it was now a newsroom too, with teams staffed like publishers. Campaigns evolved like conversations. And success came to those who could operate with cultural fluency and speed.

The SaaS Era (Software-as-a-Service). A flood of platforms promising to finally automate every corner of marketing emerged. ABM (Account-based marketing) systems, marketing automation tools, customer data platforms, personalization engines, and analytics poured into the enterprise, each offering more productivity. Every function suddenly had a tool, and every tool had a platform manager. Cloud platforms made it easier to collect data, launch campaigns, and track performance at scale. They also introduced additional complexity like new workflow dependencies and data silos. And the martech stack kept growing. In 2011, there were roughly 150 martech platforms. By 2025, that number had surpassed 15,000, according to Chiefmartec. Every new promise added another platform layer.

New roles emerged to run the machinery. As stacks expanded and tools multiplied, marketers needed people who could operate marketing technology. Roles like Marketing Technologist, Martech Architect, Campaign Operations Manager, Customer Journey Analyst, CDP Specialist, and Marketing Automation Manager emerged as critical operators behind the scenes. They were onboarding platforms, managing integrations, and maintaining data. Success now depended on teams who could design experiences but also operate the platforms that delivered them.

The AI Marketing Era (AIM). Now we're entering a new wave, and once again everything feels fundamentally different. Generative AI is accelerating content creation. Predictive intelligence is guiding decisions in real time. Agentic systems are autonomously managing tasks once handled manually. From strategy to execution, machine intelligence is embedding into every layer of marketing. It's all in the early stages, but the direction of marketing impact is undeniably clear.

The role of the marketer is about to change again.

This era is already creating new roles as well: AI Marketing Strategist, Agent Operations Manager, Marketing Systems Architect, Customer Data Engineer, Personalization Manager, Context Engineer, and Agent Governance Manager. The shift from manual execution to intelligent orchestration will require new skills like training models, designing agent workflows, engineering reasoning and decisioning, and governing autonomous systems. Traditional job descriptions will evolve, and hybrid talent across marketing and engineering will define the next wave of leadership. In this reality, the most valuable marketers will design the systems that generate and adapt in real time.

AI won't change what marketing does.
It will change how marketing works.

But here's the elephant in the room. We're not adopting AI with a clean slate. We're layering it on top of everything that came before it. On top of strategies, skills, processes, and structures that weren't designed for this kind of speed or autonomy. The opportunity is enormous, and the cost of not rethinking your operating foundation is very real.

Exhibit 2: **The Accumulated Eras of Marketing**

Era	Promise	What it Delivered	Residual Constraint
Mass Media Era	Reach	Mass audiences, iconic brands, cultural impact.	No feedback loop, centralized control, lagging measurement.
IMC Era	Consistency	Coordinated campaigns, unified voice across touchpoints.	Channel centric, calendar driven, sequential planning.
Digital Era	Precision	Addressable targeting, measurement, personalization.	Fragmented channels, rising operational complexity.
SaaS Era	Automation	Platforms, automated workflows, and data at scale.	Tool bloat, unused capabilities, new silos and costs.
AI Era	Intelligence	Real time learning, predictive decisioning, adaptive execution.	Exposes accumulated friction, requires system redesign.

Designed by Accumulation

Marketing didn't fail. It succeeded repeatedly and inherited the cost of its own success. Each wave promised new reach and new levers for growth. And each era delivered. But the old layers never disappeared. New ones were simply added on top. Most teams now operate across overlapping generations of playbooks. It's no surprise that things feel complex.

The truth few like to acknowledge is that the layers of marketing eras never went away. They accumulated.

Much of today's complexity is inherited. Marketing organizations are consolidations of skills, structures, and technologies bolted on over time. Each wave of innovation added capability but rarely replaced what came before it. The result is a patchworked function that is siloed, difficult to steer, and increasingly hard to optimize.

Ironically, marketing now has more tools than ever yet delivers only similar relative performance. Teams are data rich but often foresight poor. A Forrester report found that marketers spend 64 percent of their time on non-strategic work like switching platforms, fixing data, and pulling manual reports. CMOs live in a paradox. They oversee more teams and tools and yet have less freedom to move. Marketing may look modern, but it remains structured for the last generation of work.

The result is that marketing works harder but not necessarily smarter. Despite record investment in platforms, data, and talent, many organizations struggle to prove impact, adapt quickly, or operate with the precision that businesses now expect. This isn't a failure of effort. It's a structural problem. It shows up in siloed departments, fragmented stacks, expanding workflows, underutilized data, and misaligned teams. And now as AI enters the picture, these cracks widen. The structures that carried marketing this far won't carry us forward. What's needed now isn't another team or tool, but a new operating model.

When Adaptation Isn't Enough

Let's return to the opening conversation with the CEO. She wasn't wrong to question marketing's impact. But what she was reacting to was the strain of a structural shift. Her CMO wasn't underperforming. They were operating inside a structure that had simply outgrown itself.

Many leaders misdiagnose this problem.

They see rising costs or slow response times and assume a leadership or talent issue. It isn't. It's an architectural problem, built layer by layer over years. Marketing teams have survived by adapting. But adaptation isn't transformation. And in the AI era, this time adaptation becomes a significant liability. What's required now is a redesign.

The Future Won't Run on Yesterday's Architecture

Today, organizations are evaluating agentic systems, generative content, and predictive actions. They're designing intelligent and autonomous marketing capabilities, but we see them building them on infrastructure that was never intended to support this level of speed, coordination, or intelligence.

Costs are rising. Results are flattening. And teams are stretched thin. The symptoms include expanding departments, growing workflow dependencies, underutilized platforms, and fragmented data. Much of today's friction doesn't come from what's new. It comes from what we've retained. Marketing has adapted to every wave, but we've rarely re-engineered for how it needs to run.

The result is that teams carry the weight of generations of infrastructure. And now they're piloting AI while maintaining legacy workflows. They're running campaign calendars, managing sophisticated automation tools, wrestling with disconnected data, and pulling data reports from platforms few fully understand.

*AI's potential doesn't align with the complexity
of traditional marketing models.*

This is the reality many teams face. And yet they're asked to move faster, operate smarter, and deliver more inside systems never designed for modern marketing demands. We are living through both a performance gap and an opportunity gap. Closing either requires redefining how marketing operates.

That is why this moment demands leaders willing to ask harder questions. What are we doing simply because we always have? Where are we optimizing processes that should be eliminated? What would it take to operate with real time, machine-level intelligence?

Adapting to AI won't be enough. The cracks in marketing aren't about talent or effort. They are structural. Too often they are misdiagnosed as strategy or leadership gaps, when what's missing is the ability to translate intelligence into execution. That means platform integration, data readiness, automated workflows, agent design, and governance working together as one model. This is what modern marketing requires.

Most importantly, this isn't about replacing strategy, instinct, or creativity. It's about reinforcing them with structure, so the best ideas can be delivered at scale, with speed, and with precision. It's what marketing leaders have always wanted.

This book is for leaders ready to build that future. For CMOs who want to be known for modern outcomes. For CEOs who expect marketing to operate as a competitive advantage. And for the next generation of leaders working at the intersection of marketing and technology. This is your moment.

Summary

Every era of marketing began as progress and ended as inherited legacy layers. What once created advantage became infrastructure. What once delivered speed eventually introduced friction. Each generation of marketers adapted to new channels, platforms, and technologies, but rarely retired what came before. The result is a function built by accumulation rather than orchestration.

AI compresses the distance between demand and action. It learns continuously. It exposes inefficiencies hidden inside approvals, handoffs, fragmented data, and overlapping stacks. In doing so, it reveals a hard truth most organizations avoid. The performance limits facing marketing today aren't caused by talent gaps or effort shortfalls. They are structural. Most leadership teams still misdiagnose the problem. Rising costs, slower response, and uneven impact are treated as execution or management issues. But the root cause sits deeper. Marketing is operating inside architectures designed for a slower, more linear era.

When intelligence is treated as a feature or a tool, those limits remain. When intelligence becomes the operating fabric, the work begins to move differently. Data flows into systems that can sense context and interpret demand. Decisions happen at the right level and the right moment. Resources are allocated dynamically. Execution adapts in real time. Each outcome sharpens the next decision. That shift isn't technological. It is architectural.

Getting there requires CMOs willing to replace addition with integration, to retire work that no longer moves the market, and to redesign how marketing runs. When the operating model changes, momentum stops relying on heroics. It becomes a property of the system.

That is the purpose of this chapter and the foundation for what follows. Marketing cannot keep layering AI onto structures built for yesterday. Treat AI as the fabric of how marketing operates and rebuild the structure so intelligence can do its job.

Leadership Review

1. Are we diagnosing performance issues as leadership gaps when they are architectural failures?

2. Which legacy layers still govern how our teams work, and which are we prepared to retire?

3. Is AI embedded as our operating fabric, or bolted onto yesterday's structure?

4. Where are we optimizing processes that should be eliminated rather than automated?

5. Do we have the skills mix required to run modern marketing, including data fluency, automation fluency, model design, integration, and oversight of autonomous actions?

3

The Structural Shift Facing Today's CMOs

*The gap between how marketing works
and how it's built*

I met with the CMO of a rapidly growing data platform company. His team had just finished annual planning with new strategies, fresh ABM plays, and a retooled attribution model. On paper, it was impressive. But when I asked how confident he felt about executing it, he paused. "That's what keeps me up," he admitted. "I'm confident we have the right ideas. But getting them out the door as planned, across functions, with our resources…yeah, that's where we'll lose momentum."

The problem wasn't his team. It was orchestration. "We're operating in fragments," he said. "And we're burning time stitching it all together." The conversation didn't end with adding more resources. It ended with new ways to structure the operating model. Because in today's environment, even the best strategy without modern execution is table stakes.

We know that markets today move at a velocity that most marketing teams aren't designed to handle. Buying signals are tracked in real time. LLM systems (e.g., ChatGPT, Claude, Gemini, xAI) generate complete answers. Actions unfold continuously across channels. No planning calendar, manual process, or organizational sprint can keep up. This is a structural break, and AI is the catalyst forcing CMOs to confront it.

Chapter 2 showed how decades of innovation layered tools and channels onto marketing while compounding complexity. This scaffolding is now pushing the traditional model beyond its limits. Today, marketing momentum can no longer be managed with legacy structures. The future must be engineered.

History offers precedent. Television created scale. The internet created access. Digital created precision. Mobile created immediacy. Each wave reshaped competition by changing how marketers reached customers. AI is different. It changes *how* marketing runs. And it changes *how* customers behave. It redefines the fundamental motions of marketing, including how demand is sensed, how decisions are made, and how action happens in real time.

AI isn't another marketing wave. It's a structural game changer.

Automation has become operational. Intelligence flows are embedded in daily work. Martech platforms have become execution engines. Intel's Andy Grove famously described moments like this as an inflection point. A juncture when the fundamentals of a business are irreversibly forced to change. This shift reflects the advance of AI, but also the reality that marketing is now deeply data and technology driven, while customers increasingly rely on AI to guide decisions. Together, these forces create the break.

Marketing is crossing a threshold. What once moved in slow, sequential bursts is now expected to operate as a continuous, adaptive system. Campaigns built on static messages give way to operating models driven by intelligence. Momentum no longer comes from isolated moments, but from compounding performance over time.

Most marketing organizations were never built for this shift.

CMOs now face a clear choice. Treat AI as a productivity layer on top of traditional marketing, or redesign marketing to engineer intelligent performance. Modern marketing is no longer defined by what you run. It's defined by what you build to sense, decide, and act faster than any legacy model can. The current operating model has expired.

From Moments to Momentum

At the start of every fiscal year, a retail CMO hosted her annual planning meeting to line up holiday campaigns and seasonal pushes. The team's energy revolved around those spikes. Every milestone was a "moment" designed to capture attention, intercept intent, and drive a lift in revenue. She confided to me that the lift didn't last. Sales would peak, then flatten. Momentum slipped away and then had to be rebuilt. Across the industry, this approach was considered normal. But now AI is unpacking new levels of buying intent.

Marketing has always prized the "moment" of the right message delivered to the right person at the right time. For decades, traditional marketing ran on a campaign centric model that was linear, predictable, and largely manual. It ran on planning cycles and broad targeting, designed for audiences who consumed passively and responded slowly. Work was resource intensive. Processes were loose. Insights arrived after the fact, if at all. The focus was the moment, including launch dates, scheduled posts, media buys, news releases, and event calendars.

This traditional logic no longer applies. Today, intelligence moves in real time. Customers expect personalization, interaction, and relevant content instantly. A consumer who gets meals delivered, streams on demand, scroll curated feeds, or orders a ride within seconds has little patience for a marketing experience that takes days to align content or launch an offer. AI and automation have closed that cycle. Data fuels execution as it happens. Workflows orchestrate themselves. Systems adapt in motion. It's a different model built on momentum, and the dynamics of this shift are easy to compare.

Exhibit 3: **The Shift From Moments to Momentum**

Traditional Marketing	Modern Marketing
Campaigns built around *moments*	Systems engineered for *momentum*
Episodic launches	Continuous flow
Manual workflows	Automated orchestration
Static targeting	Dynamic personalization
Siloed data	Real time signal integration
Reactive insights	Predictive intelligence
Martech accumulation	Martech integration
Marketing departments	Marketing orchestration

Campaigns built around moments can still generate bursts of visibility, but they rarely deliver lasting advantage without sustained investment. Organizations that cling to this model spend heavily just to intercept attention that disappears as quickly as it arrives. But with modern AI enabled marketing momentum, each interaction strengthens the next. Each decision sharpens future actions. Each cycle expands capacity. Instead of chasing audiences, modern marketing moves with them by adapting to behavior shifts, and learning from every touchpoint.

Marketing momentum must be engineered. The companies thriving today realize the speed of AI is outpacing the speed of traditional marketing decisions. *They are building systems that sense, interpret, decide, allocate, activate, and learn at market speed.* The technology already exists to make this possible. The question is no longer whether it can be done. The question is how CMOs will get it done.

This is the inflection. From campaign logic to system logic. From isolated wins to engineered performance. From manual execution to autonomous flow. From moments to momentum. Collectively, this is likely to be the most significant shift in marketing's history.

A Strategic Shift, Not an Optional Upgrade

Martech platforms, data ecosystems, and AI are no longer differentiators. The baseline infrastructure of many parts of marketing is already automated, and data driven. Technology is at the center. The goal now is to build the systems that scale intelligence, accelerate speed, and drive continuous learning in real time. This isn't another wave of "digital transformation." Most functions already believe they are digital, and in many ways they are. The real change is applying the expertise to design *how* marketing runs across a modern system, structure, and operating model. That is where the next generation of market leaders will compete. And this is how marketing now creates a competitive edge.

What separates leaders from laggards now isn't the size of the stack but the design of the system.

This section defines how teams collaborate, how data feeds decisions, how platforms interconnect, and how technology integration ladders to modern performance. It's more than productivity and efficiency. It's about designing a system where the right actions happen intelligently and automatically.

What an intelligent operating model enables:

- Automation and AI driven orchestration as the way of working

- Data platforms and ecosystems feeding decisions in real time

- Integrated martech platforms connected as one system, versus scattered tools

- Agentic AI embedded into workflows, automating and adapting actions

- Generative AI producing tactical variations at scale and speed

- Analytics and telemetry that measure system performance

This is the shift between activity and structural advantage. For the first time in decades, marketing is redefining how it runs. This is where advanced technology meets the deep functional expertise CMOs have spent careers building. Creative and code, brand and performance, art and algorithm are fusing into a function designed at scale.

As this shift accelerates, the question facing every CMO is no longer whether to adapt to AI, but how quickly they can redesign the operating model. Achieving momentum can't be left to chance. After years inside marketing organizations, I've learned it's a design choice. In this modern era, advantage doesn't belong to the strongest brand or the biggest budget. It belongs to the teams that move with precision and executes with intelligence.

Today's operating dynamics repeat across industries. Strong strategies, capable teams, and full martech stacks are common, yet scale and momentum slip away. The issue isn't lack of effort, it's about outdated operating models. When legacy workflows and processes cannot match the speed of the market, complexity takes over. Marketing teams know them as missed deadlines, redundant reviews, bloated martech stacks, and decisions that arrive too late. They drain energy, inflate cost, and dilute impact because the operating model hasn't evolved to meet the ambition of the strategy and plan.

This is why marketing leaders must honestly examine how work actually happens. Where workflows stall. Where talent is underused. Where integration breaks. Where platforms sit idle. Where approvals slow progress. These are structural vulnerabilities. And they determine whether marketing remains a cost center or reclaims its role as a growth engine in the AI era.

The Silent Killer of Momentum

From the outside, most marketing organizations look busy. Campaigns are in motion. Content is in development. Platforms are in use. Yet performance growth is linear at best. Speed breaks down, alignment struggles, and growth targets slip. What appears to be performance issues is increasingly becoming operating issues.

Friction is the silent killer of momentum. It enters through outdated workflows, underused resources, and decision bottlenecks. It doesn't show up on dashboards, but leaders feel it in missed launches, stalled personalization, and opportunities lost in approval chains. CFOs see it in rising costs. CMOs feel it in missed forecasts. CEOs sense it in the widening gap between ambitious strategies and uneven results.

This isn't a failure of effort. It's a failure of structure.

The cost of latency shows up as wasted time and lost growth. It is both operational and cultural. Teams burn out doing work that feels harder than it should. Leaders lose confidence. Innovation gives way to maintenance. The pressure CMOs feel rarely comes from a lack of strategy or creativity. It comes from stretched resources and fragmented alignment.

Operating friction is structural, and it shows up in five recurring pressure points:

1. **Decision latency**: Markets move in days, but approvals take weeks. Programs don't stall because strategy is wrong, they stall because teams struggle to move them forward. Creative needs sign off. Legal needs reviews. Sales needs enablement. By the time approval arrives, the moment has passed. *Snapshot*: A seasonal promotion for a consumer product required corporate, legal, and management reviews across four markets. What should have launched in 48 hours took ten days. Competitors owned the opportunity before the campaign went live.

2. **Platform sprawl**: More tools create more silos, more logins, and more governance overhead. In our work, we see the average enterprise stack runs 50 to 100 platforms, with often some overlap. Instead of acceleration, teams inherit complexity. *Snapshot*: A regional bank operated twelve tools for CRM, email, personalization, and analytics. New campaigns required three days just to stitch customer data into outreach. Despite the new platforms, conversion rates ran consistently below forecast. We've also identified that every redundant tool adds five to eight percent of overhead in licensing, training, and maintenance.

3. **Signal lag**: Data is everywhere, but insight arrives late and fragmented. Decisions are made on what happened last week or last quarter, not what's unfolding now. In the AI era, timing is leverage. *Snapshot*: At a B2B SaaS provider, customer data was manually exported between automation platforms. By the time campaigns launched, fifteen percent of the contacts had decayed. Performance was dropping before campaigns launched. When signals lag, competitors act on today's opportunity while you chase yesterday's story.

4. **Workflow friction**: Every handoff is a point of lost momentum. Rewrites, redundant reviews, and endless meetings make execution highly variable. Without designed workflows, every project becomes an exception that doesn't scale. *Snapshot*: A CPG brand's campaign brief required five team touchpoints before production. Creative talent spent more time in alignment calls than on design. High performers burn out when their skills are spent navigating bottlenecks instead of creating value.

5. **Talent drain**: When operating models fail, people compensate. Executives chase approvals. Creatives focus on revisions. Channel managers try to integrate. Regional teams await guidance. Over time, drag reduces high performance. *Snapshot*: A CMO described his top analyst as "half data scientist, half firefighter." Instead of building predictive models, he spent seventy percent of his time

troubleshooting data integrations. Talent leaves when their job becomes maintenance instead of growth.

Individually, these examples slow execution. But together, they act as a compounding tax on marketing impact. Every delay, layer of complexity, and wasted handoff drains energy and erodes impact. The cost is financial, cultural, and competitive.

Treat operating latency as a marketing tax.

These nuances of everyday marketing execution are costly. And their impact will only be magnified against the backdrop of AI. They never appear on P&L, yet it's paid every day in lost growth, wasted time, and diminished credibility. And sustained effort cannot overcome it. The harder teams push, the heavier it becomes. What looks like a performance problem is a structural design problem. For CMOs, the implication in the AI era is that speed can't be fixed with willpower. The good news is this latency can be engineered out. Because no amount of talent, budget, or ambition can sustain momentum in the AI era.

From Effort to Engine

A CMO at a scaling SaaS firm once told me, "We don't lack ambition, but we may lack weeks." Her team was talented and motivated, yet every launch required heroic efforts. Priorities piled up. Meetings multiplied. The harder they pushed, the slower they ran. "Some days it just feels like survival," she said. This is the risk of effort.

For years, marketing compensated for structural limits with hustle. A lot of it. Teams worked late, routed around broken workflows, and forced campaigns across the finish line through sheer determination. In the old model, that hustle sometimes created an edge. In today's environment, it creates exhaustion.

Momentum isn't the product of hustle. It's the outcome of system architecture.

Momentum now requires a shift from effort to system. Speed no longer comes from adrenaline. It comes from design. In practice, here is what this looks like:

- Systems that act without waiting, as workflows trigger on signals rather than managers

- Work that adapts automatically, with offers, content, and journeys optimizing in real time rather than quarterly cycles

- Decision authority embedded in the work, reducing approvals, delays, and meetings about meetings

- Clearer flow, with less rework, duplication, and handoffs

This is what a well-designed operating model makes possible. It aligns data, simplifies platform management, and integrates intelligent orchestration so output becomes predictable and scalable. Momentum stops depending on heroics and starts becoming the natural state of the modern marketing system.

Exhibit 4: **From Manual Effort to Engineered Flow**

Where Friction Shows Up	**What Manual Effort Does**	**What Engineered Flow Enables**
Decision Latency	Leaders chase approvals manually.	Decisions embedded into workflow logic.
Tool Sprawl	Teams toggle across platforms.	Unified stack with active portfolio management.
Signal Delay	Analysts scramble to update reports.	Real time signals feeding predictive models.
Workflow Crawl	Managers coordinate handoffs by hand.	Automated orchestration, minimal exceptions.
Talent Drain	High performers cover structural gaps.	Talent redirected to creativity and strategy.

High performing teams in this model aren't doing more. They are doing less with greater clarity, sharper alignment, and far fewer points of friction. They move as systems, rather than collections of individuals and functions. The organizations that thrive in the AI era will be those that stop treating speed as a scramble and start treating it as a design principle.

The Shift in Practice

Organizations across industries are already showing what happens when latency is engineered out of the system and AI enabled momentum is built in.

A global tech company predicts pipeline gaps before they happen. For years, the company relied on quarterly pipeline targets to understand whether marketing was meeting sales expectations. By the time those reports arrived, it was already too late to course correct. Campaigns were in market, budgets were spent, and the sales team was left scrambling to make up the difference. These were insights arriving long after action was possible.

By embedding predictive models directly into campaign planning, the company rewired the process. AI agents synthesized intent data, CRM velocity, and content performance daily. Gaps that previously took months to surface were flagged in near real time. Marketers adjusted campaigns on the fly, reallocating spend, refining content, and shifting priorities before opportunities were lost.

A wellness brand reclaims thirty percent of team time with autonomous operations. This fast-growing wellness brand had built a loyal customer base, but scaling its marketing came at a cost. Launch cycles stretched to three weeks. Every campaign required long checklists of manual steps like assigning tasks, tracking dependencies, and producing content variants by hand. The highly manual workflow consumed the team's energy.

By introducing an automation layer, execution began to run differently. Tasks were assigned autonomously. Blockers were flagged instantly. Content variants deployed dynamically based on live engagement data. The cycle time fell from almost three weeks to five days. We saw about thirty percent of team time was reclaimed. Talent shifted to strategic partnerships. Modernizing the workflows accelerated output and elevated quality.

A financial services firm turns martech chaos into clarity. A CMO we had worked with took a new role at a financial services company. He discovered eighty-seven martech platforms across marketing and sales. Teams were overwhelmed with coverage, logins, duplicative processes, and overlapping features.

Through active portfolio management, the stack was simplified. Twelve redundant tools were retired. The remaining platforms aligned to a universal data layer, ensuring campaigns drew from consistent, real time information. Execution no longer required patchwork integrations. It flowed through a connected architecture. The difference was immediate. The CFO, who once viewed the stack as wasteful, began to see marketing technology as a strategic asset. ROI reporting became credible and team collaboration improved significantly.

A consumer brand rewrites journeys in real time. This consumer brand had invested heavily in persona-based journeys. Dozens of profiles were carefully mapped with branching campaigns for each. But by the time journeys launched, they were already outdated. Customers moved faster than the personas could adapt. The static design was too rigid for real time behavior.

Machine learning changed the work. Instead of persona tracks, the system shifted to next best action modeling. Journeys became dynamic, adapting continuously to customer signals. Offers, messages, and touchpoints evolved as behaviors shifted, without waiting for new segment briefs. Results were striking. Engagement lifted four times. Conversions rose twenty two percent. More than numbers, the brand's marketing felt alive as it evolved with their customers.

These examples are proof points of a larger truth. The organizations pulling ahead aren't simply better staffed or better funded. They are better designed. Momentum comes from orchestrated systems built to sense, interpret, decide, allocate, activate, and learn without latency. This is modern marketing engineered.

Leadership Commitment

Every inflection point in marketing history has required strong, committed leadership. Television, digital, and social each forced executives to expand how marketing engaged customers. Each shift required new skills and new partners. But AI is different. It resets the operating model. That responsibility cannot be delegated.

Most teams today operate on structures never designed for real time, AI driven growth. They rely on extraordinary effort to bridge gaps by chasing approvals, stitching data, and patching martech integrations. Too often, progress depends on brute force to cross the finish line. We've all been there. Over time, this approach creates fatigue, inflates cost, and fails to scale with growth. It's simply not sustainable.

In the AI era, while you coordinate, competitors accelerate.

Artificial intelligence makes teams better. When people get better at their jobs, organizations want more from them, not less. This shift requires leaders who can imagine what is possible and prioritize it. There will always be launches to deliver, reviews to attend, and fires to fight. But unless the system becomes the priority, surface AI efforts will deliver only marginal returns. Momentum isn't something to delegate. It's something to design and to lead.

For CMOs, building modern marketing momentum is now central to the role. It means exposing and eliminating failure points with the same urgency as a failed campaign. Workflow bottlenecks, platform sprawl, and data lag aren't minor nuisances. They are structural risks. CEOs share responsibility for reframing marketing from a cost center to a growth system, from a busy function to a compounding engine. No CEO would tolerate an obsolete supply chain or an outdated manufacturing line. And no CEO can afford a marketing structure built for another era.

Leaders who embrace this responsibility will reset its strategic relevance inside the enterprise. Those who delay will see influence, budgets, and trust erode. The question is no longer whether change is necessary. It is how quickly you will move to deliver it.

If you lead business or marketing today, IBM's e-business shift offers a blueprint for transformation. When foundational technology changes, you must redesign the operating model. Treating AI as just a tool will leave you playing catch-up. Designing for transformation allows you to lead.

IBM's Shift to e-Business

During the mid-1990s, IBM made one of the most consequential strategic pivots in corporate history. Under CEO Lou Gerstner, the company recognized that the internet wouldn't simply enhance business. It would redefine it. IBM's ambition shifted from selling hardware and software to helping the world's largest enterprises reinvent themselves as network-centric businesses. That transformation became known as e-business.

As Director of IBM Corporate Marketing, I had the privilege of helping shape and lead that shift. Our IBM and O&M team launched the global "e-business" campaign to redefine IBM's relevance in the new internet era. The internet was largely perceived to be a consumer phenomenon. The goal was to show how IBM services and products help companies do real business on the web. The campaign became one of the most successful in IBM's history, earning Adweek's Campaign of the Year and an AMA Gold Effie for marketing effectiveness.

Beyond the global brand platform, we built IBM's first cross-enterprise e-business demand program and one of the largest early display campaigns on the web. We created a unified message system, partnered with business-unit CMOs, and aligned IBM's story around a single enterprise narrative of e-business reinvention.

Inside IBM, this work made a deeper point visible. Technology products alone weren't enough. Advantage came from the integration of technology, services, process, and brand moving together. That conviction still shapes how I see modern marketing. Every era reaches its inflection. For IBM, it was e-business. For today's leaders, artificial intelligence is the next operating fabric of marketing. Treat the AI shift as foundational, redesign how the function runs, and lead the change from the inside.

Summary

There was a moment when the internet stopped being a playground and became the baseline for how business and marketing worked. It completely transformed how customers discovered, decided, and purchased. Category leaders across industries changed rapidly, and new brands emerged. Leaders who recognized the shift early redesigned how they operated and never looked back. The ones that treated e-commerce as a side project were disrupted. The case studies are abundant.

We are at that moment again. AI is quickly moving from pilot to mainstream. It collapses the lag between insight and action. It reads context at a scale. It learns from every outcome and carries that learning into the next decision without waiting for a meeting. If you treat AI as another tool, progress feels incremental. But treat AI as the foundation and the work begins to move on its own.

This chapter has made the case in operational terms. Campaign logic creates spikes that fade. System logic creates momentum that compounds. And making this shift is a structural choice. Decision latency, tool sprawl, signal lag, workflow drag, and talent waste drain performance even when strategy is strong. The organizations that pull ahead run decisions and actions in real time. The ones that fall behind still wait for cycles and status meetings.

The leadership decision is unavoidable. Accept that AI isn't an accessory. It's the operating fabric that will govern how marketing works. Proof must shift from activity to advantage. Show that investments are moving from incremental efficiency to structural performance. And give authority to the people who translate intelligence into execution and remove the approvals that slow them down.

The risk is direct. If AI is treated as incremental, marketing will be bypassed by environments that already know what customers want. If intelligence is treated as foundational, the function resets its role in the enterprise. Work becomes continuous. Handoffs decline. Surprises at the end of the quarter diminish. Results strengthen and then strengthen again. Do you want to advance or be a disruption case study?

Leadership Review

1. Are we still running marketing on effort and coordination, or redesigning it to move on its own?

2. Do we treat AI as another layer of automation, or as the new operating fabric of how marketing works?

3. Do leaders spend more time managing activity, or designing systems that continuously produce outcomes?

4. Where does latency still slow decisions, delay action, or drain talent?

5. When the next board review arrives, will we present more activity, or a marketing function with intelligence built in?

4

The AI Marketing Operating System

*Introducing the foundation for
how modern marketing operates*

Recently, I asked a CMO what his leadership team was doing with AI. He smiled, then shook his head. "Well, it depends on who you ask," he said.

He went on to tell me that one leader was running a personalization pilot but couldn't move beyond small test groups. Another had launched a Generative AI effort to produce regional tactics, only to have the brand team resist because it was off brand. A third was working with a vendor promising predictive models, but sales didn't trust the recommendations.

"Overall, I would say some of my team is excited. They're convinced AI will unlock innovation," he continued. "Others are skeptical. A few others are nervous it will automate them out of a job. And I've heard some are running shadow projects. So yes, it's all over the place and it's becoming a real distraction."

He paused before continuing "But to be honest, if my CEO asked me today what our AI strategy is," he said, "I'd struggle to give a clear answer. We don't have a destination. We're still experimenting, and it hasn't changed how we run."

That conversation sounded familiar because the words he used mirrored other discussions. Statements like *skeptical teams, pilots and experiments, shadow usage, wait and see, using some AI features on platforms, data still needs work, proof of concept mode, watching what competitors are doing, and waiting for IT.* All of these phrases are indicators that teams are trying to use AI, but lack a collective point of view on what they're working towards.

And what's happening is even more fundamental. Teams are running next generation AI capabilities on top of legacy operating models. The mismatch is structural, and it becomes more visible as AI accelerates. These leaders can see the inflection point, but they lack a path for crossing it. This chapter starts from that recognition. Progress doesn't come from more experiments. It comes from building a new marketing foundation.

This foundation is the *AI Marketing Operating System (AIM OS).*

Engineering Marketing Performance

For decades, marketing performance was defined by effort. Bigger teams, larger budgets, and more programs were assumed to produce greater impact. If you could outspend or outstaff competitors, then relative momentum followed. That model worked in slower eras, when channels evolved gradually and customer expectations shifted at a manageable pace. That environment no longer exists.

What has changed isn't simply technology, but how customers make decisions. Intelligence now shapes buying experiences directly. Customers expect interactions that reduce the effort of choosing. They expect brands to understand context, guide decisions, and remove complexity across every touchpoint. These expectations are already influencing how products evolve, how experiences operate, and how trust is formed.

So as buying changes, the limits of effort-based marketing become more visible. Smarter experiences now deliver deeper information and faster answers. In this environment, intuition, drive, and scale alone cannot keep pace. Operating models built on labor, approvals, and sequential workflows move way too slowly for markets shaped by real time intelligence.

At the same time, marketing influence has become more distributed. Discovery, comparison, and shortlisting increasingly occur inside consumer models that summarize options and recommend outcomes are already emerging across AI powered search and assistant platforms. Marketing doesn't lose relevance in this shift, but its leverage point shifts. Performance still depends less on managing individual interactions and more on shaping how the brand is represented, evaluated, and trusted across AI mediated environments.

These changes redefine what performance means. Marketing is no longer measured by how many assets are produced or how many campaigns are launched. It's measured by how effectively an organization can sense demand, interpret context, decide priorities, allocate resources, and act in motion. The ability to operate this way determines whether marketing can meet modern consumer expectations with relevance and speed.

This is the gap the AI Marketing Operating System is designed to close. AIM OS provides the structure that allows marketing to function as an intelligent operating model rather than a collection of activities. It embeds intelligence into decision making, execution, and learning so performance improves continuously.

*In this shift, AI reframes marketing
from activity to operating model.*

Functional performance now depends on how quickly consumer intelligence becomes action, how seamlessly decisions move into execution, how precisely activity aligns to customer intent, and how consistently the organization learns in real time. The structural difference between effort-driven marketing and engineered performance is summarized below.

Exhibit 5: **The Structural Shift in Marketing Performance**

Performance Driver	Then	Now
Core Input	Individual effort, team capacity	System design, standards, automation, AI
Campaign Success	Reach, impressions, budget allocation	Throughput, speed, signal driven impact
Execution Model	Manual, sequential	Automated, responsive, always on
Optimization Cycle	Monthly or quarterly reviews	Real time feedback systems
Martech Value	Individual feature sets	Integrated system behavior
Role of Technology	Supportive infrastructure	Central operating system (AIM OS)
Leadership Focus	Managing activity	Designing and governing the operating system
Measurement Focus	Channel metrics	System performance and business outcomes

The discipline moves from effort driven outreach to system driven flow. From rigid campaign cycles to adaptive feedback systems. From reactive optimization to proactive intelligence. More importantly, technology has moved from the background support to the foreground of performance. Platforms that once automated activity have become the architecture through which marketing drives execution. Data shifts from explaining the past to powering what happens next. Artificial intelligence moves from experimentation to guiding decisions in real time.

This shift creates a new mandate for marketing leaders.

The CMO moves from orchestrating loosely connected activities to owning the operating model. The role is to architect and iteratively evolve the operating system that delivers next generation performance.

The leadership challenge shifts from doing more work to engineering better outcomes. In an era of real time everything, performance scales through AI driven system design, not additional exertion. This is the shift. And the market winners will be the teams whose operating systems are already in motion. The remaining question is whether you will build yours before your competitors do.

AI Marketing Operating System (AIM OS)

What modern marketing lacks isn't more strategy, planning, or ambition. It lacks an operating model designed for intelligence in motion. The **AI Marketing Operating System (AIM OS)** is that model.

AIM OS is the structural foundation that allows marketing to operate as a coherent, intelligent system rather than a collection of disconnected activities. As mentioned earlier, it brings data, platforms, automation, and artificial intelligence into a single operating flow.

This isn't another technology stack. It's how modern marketing works.

As customer behavior accelerates and influence spreads across environments marketing doesn't control, performance depends less on individual programs and more on how intelligence moves through the organization. AIM OS makes that movement possible. It is the operating layer that turns insight into action, action into learning, and learning into sustained advantage.

At the core of AIM OS are *four performance capability zones*. These zones are not optional features or modular add-ons. Each represents a fundamental function marketing must now perform. Individually, they create lift. Together, they define modern marketing performance.

The Four Performance Capability Zones of AIM OS

1. **Adaptive Decisioning**: The capability to continuously sense signals, interpret context, and determine next best actions in motion. Without it, marketing remains anchored to hindsight and periodic planning.

2. **Autonomous Activation**: The capability to execute decisions instantly and at scale across channels. Without it, execution latency becomes the hidden tax on strategy.

3. **Experience Loop Intelligence**: The capability to learn from every interaction and feed that learning back into the system. Without it, marketing optimizes assumptions instead of behavior.

4. **Intelligence Governance**: The capability to guide, constrain, and make intelligence trustworthy at scale. Without it, speed creates risk and automation erodes trust.

These four zones are engineered to work as one integrated system. No single zone delivers modern performance on its own. Adaptive Decisioning without Activation stalls. Activation without Learning wastes effort. Learning without Governance creates risk. Governance without Decisioning slows progress.

AIM OS exists to fuse these capabilities into a continuous operating flow.

This is the operating model that allows marketing to keep pace with markets shaped by intelligence. And it's the structure the rest of this book will build on.

Exhibit 6: **The AI Marketing Operating System**

Capability Zone	Primary Purpose	What It Enables
Adaptive Decisioning	Intelligence and reasoning	Interprets signals, applies models, determines next best actions in motion
Autonomous Activation	Execution at scale	Delivers content, offers, and experiences automatically as decisions are made
Experience Loop Intelligence	Continuous learning	Captures interaction feedback to refine decisions and improve performance over time
Intelligence Governance	Control and accountability	Sets guardrails, standards, and oversight to ensure trust, alignment, and compliance

In this model, AI functions as operational intelligence embedded into day-to-day execution. Decisions accelerate. Precision increases. Performance becomes something leaders actively manage, versus something they review after the fact. This marks a step change in how marketing operates and scales.

Modern marketing runs on AIM OS.

AIM OS enables structural momentum. When the operating system underneath is engineered, the experience remains customer-focused and expressive. As signals, models, logic, and governance move together, performance shifts from episodic to continuous. Each cycle sharpens the next. Organizations that understand this shift know to stop asking whether marketing can deliver on all fronts. Instead, they see their AIM OS operating at the speed of the market.

The sections that follow examine each performance capability zone, showing how they operate individually and how, together, they define modern marketing performance. The four zones are as follows:

Zone 1: Adaptive Decisioning (Intelligence and Reasoning)

Every operating system requires a decision layer. In modern marketing, that layer is Adaptive Decisioning. It provides the ability to sense signals in motion, evaluate options against models and rules, and determine next best actions with speed and precision.

Historically, marketing decisions moved on planning cycles. Teams analyzed prior-quarter data, debated direction, and set priorities, messages, and tactics into motion based on what had already happened. By the time programs reached the market, conditions had shifted. Customer behavior evolved and competitors adjusted.

Adaptive Decisioning changes the clock speed of marketing. Decisions move from periodic to continuous. Instead of waiting for reviews, the system interprets real time data streams as they arrive. Instead of reacting after the fact, intelligence operates in the moment against inputs including who visited, who engaged, who downloaded, who registered, who abandoned, who reached out. Models evaluate those signals, with other data sets to determine what should happen next. This capability is built on three engineered elements working together:

- *Signals*: Data flowing from every channel and interaction, structured and unstructured, behavioral and contextual.

- *Models*: Predictive and generative AI that interpret signals, project outcomes, and weigh alternatives.

- *Logic*: Orchestration frameworks that translate model output into action: who to engage, what to deliver, when to act, and where interaction should occur.

When these elements operate as one, marketing gains a different kind of agility. A promotion that underperforms in the morning can be recalibrated by the afternoon. A customer signaling intent in one channel can be invited to a relevant experience in another, immediately. Campaigns stop moving in rigid bursts and begin adapting with market behavior.

Many marketing teams already possess elements of this capability. This includes a recommendation engine in ecommerce, or a lead scoring model in account-based marketing. But components alone don't create intelligence. Without integration, each operates in isolation, blind to broader context. Adaptive Decisioning delivers lift only when signals, models, and logic are engineered into a unified decision layer.

This introduces a shift in leadership responsibility. The CMO still sets direction, allocates investment, and aligns teams. But now the role expands to oversight of the decision layer. Leaders must ensure data flows are complete, models are trained and governed, and orchestration logic reflects customer needs, business priorities, and operating standards. This replaces intuition. It's marketing engineered to think in motion. And it forms the foundation on which every other zone of AIM OS depends.

Zone 2: Autonomous Activation (Execution at Scale)

Decisions create value only when they move into action. Autonomous Activation is the execution layer of AIM OS. It converts intelligence into action instantly, at scale, and across channels, without waiting for constant manual intervention.

In traditional marketing, execution has long been the constraint. Creative assets required approval cycles. Campaign calendars dictated timing. Channel teams queued tactics. Regional and business line teams struggled to keep pace. Even strong strategies stalled in handoffs, resourcing gaps, and operational delays.

Autonomous Activation removes that friction.

The system delivers the right message, in the right channel, at the right moment, directly from decision logic. Content, offers, and experiences activate automatically, aligned to real time signals and continuously refined through feedback. Execution becomes a property of the system, rather than a coordination exercise for teams.

This capability is built on three layers working together:

- *Content Readiness*: Modular, tagged, machine readable assets that can be assembled and deployed dynamically.

- *Channel Connectivity*: APIs and orchestration frameworks that link advertising, email, web, mobile, social, and events into a shared execution fabric.

- *AI Orchestration*: Models that determine how actions unfold across channels, sequencing delivery in ways that feel coherent and on brand to the customer.

When these layers operate as one, execution accelerates. A customer exploring a product online can receive relevant content immediately, transition seamlessly to another channel, and engage with personalized offers in context, without reliance on static journey maps or manual execution.

Most organizations experience early versions of this through basic automation such as triggered emails, retargeted media, or chatbot responses. These fragments scale volume, but they rarely coordinate across channels or adapt in context. Autonomous Activation operates differently. It executes holistically, guided by live decisioning and continuous learning.

For CMOs, this changes the leadership mandate. Attention shifts from approving individual executions to designing a system that is trustworthy, brand safe, and outcome aligned. Guardrails, governance, and creative standards move upstream, embedded directly into how automation operates.

With Autonomous Activation, marketing moves from calendar driven pushes and static journeys to a continuous flow of intelligent, customer facing actions. Speed increases. Relevance deepens. Execution evolves from a bottleneck into a durable competitive advantage.

Zone 3: Experience Loop Intelligence (Continuous Learning)

If Adaptive Decisioning provides reasoning and Autonomous Activation delivers execution, Experience Loop Intelligence supplies learning. It ensures that every customer interaction feeds back into the operating system, sharpening relevance, speed, and performance with each pass.

Historically, marketing treated execution as a one-way push. Messages were designed, launched, and reviewed later. Measurement arrived weeks or months after action, often distilled into static reports. Success was defined by what went out, and rarely what came back. Experience Loop Intelligence reverses that dynamic.

Every interaction becomes a signal. Responses and non-responses generate data. Engagement patterns update what the system understands about individuals, segments, and markets. Learning no longer waits for quarterly reviews. It happens in the flow of engagement. This capability is built on:

- *Collection*: Instrumentation across channels and journeys that captures granular interaction data, from explicit actions like purchases to subtle signals such as dwell time or engagement.

- *Interpretation*: Models that separate signal from noise, surface patterns, and identify the drivers teams may not detect on their own.

- *Application*: Mechanisms that feed learning back into decisioning and activation, so the next action reflects what was just learned.

When these layers operate as one, marketing begins to behave like a learning system. Messages that resonate with a specific audience can be amplified immediately. Offers that underperform can be withdrawn before budget is consumed. Over time, behavioral patterns emerge that inform strategy at the portfolio level, versus just within individual programs.

Most organizations already have fragments of this capability. Dashboards. A/B tests. Surveys. Attribution models. But these tools tend to be slow, disconnected, and backward-looking. Experience Loop Intelligence requires engineered integration that makes learning immediate, continuous, and actionable.

For leaders, the implication is structural. Marketing still manages journeys, but responsibility expands to designing a system that listens as effectively as it speaks. A system that learns from the market and becomes more precise with every interaction. With Experience Loop Intelligence in place, marketing moves beyond broadcasting campaigns. It operates as a dynamic, always-learning exchange with the market, advancing continuously rather than waiting for the next review cycle.

Zone 4: Intelligence Governance (Control, Trust, and Accountability)

Every high-velocity system requires control. Within AIM OS, that control is Intelligence Governance. This capability ensures marketing intelligence remains transparent, responsible, and aligned with business priorities. It allows speed and automation to scale without compromising trust.

Historically, governance in marketing focused on brand guidelines, privacy compliance, social policies, and budget oversight. These controls mattered, but they often operated after the fact. Reviews followed launches. Audits followed issues. Rules were applied as patches rather than built in as foundations.

Intelligence Governance operates differently. Governance is engineered into the operating system from the start. Guardrails, ownership, and visibility are designed directly into how decisions are made and actions are executed. This keeps AI driven marketing aligned as it learns and adapts, ensuring performance advances within clearly defined boundaries.

This capability rests on three interdependent pillars:

- *Data Integrity*: Inputs are accurate, accessible, and privacy compliant, allowing models to learn from trusted sources and reducing bias risk.

- *Model Oversight*: Algorithms are monitored for performance drift, decision patterns are visible, and human judgment remains available where nuance matters.

- *Policy Orchestration*: Brand standards, regulatory requirements, and ethical principles are embedded directly into AIM OS, so compliance becomes a design feature rather than a downstream check.

When these pillars are in place, marketing gains a rare advantage: speed with safety. Personalization scales without crossing ethical lines. Campaigns adapt in real time while respecting privacy. Automation runs continuously within brand and regulatory boundaries.

For CMOs, this zone defines a critical leadership responsibility. Performance and governance now move together. Leaders must steward how intelligence is used, ensuring transparency to boards, accountability to regulators, and trust with customers.

Intelligence Governance allows AIM OS to scale responsibly. As marketing moves faster, it also moves with greater control. And in doing so, it positions marketing as a trusted architect of how the enterprise applies intelligence in the market.

The Power of AIM OS

AIM OS brings real time intelligence and precision to everything marketing touches, from product launches and promotional timing to brand activation, customer engagement, and cross functional collaboration. It enables marketing to adjust in motion, personalize experiences at scale, respond to sentiment as it forms, and align activity with measurable clarity.

The leverage comes from how people and intelligence work together. AIM OS is most powerful when strategic thinkers, creative leaders, marketing managers, and engineers operate as one. With the right structure in place, judgment moves up the value chain. Teams spend less time coordinating execution and more time shaping brand platforms, customer relevance, competitive advantage, strategic timing, and market momentum.

Marketing gains more than speed. It gains clarity, agility, and a direct connection to growth. With the right metrics, leaders manage performance structurally and repeatably, rather than tracking output in isolation. In modern marketing, the surface remains creative, but the operating system underneath is engineered. This is the AI marketing mindset shift.

Creativity, brand instinct, and strategic judgment remain central. AIM OS amplifies them by embedding intelligence directly into how marketing operates. Advantage comes from how well the system performs. Leaders move from asking what to launch to understanding how the system launches it. From counting outputs to engineering the conditions that produce better outcomes.

Adopting an AI marketing mindset means thinking in systems rather than silos. It means understanding how data flows, how decisions move in motion, and how learning compounds at scale. And it means leaving behind calendar driven moments in favor of continuous orchestration, where strategy, content, and engagement evolve dynamically in real time.

Applying AIM OS Across B2B and B2C

Whether marketing serves businesses or consumers, the operating principles are the same: sense demand in real time, decide faster, automate execution, and govern with confidence. What differs is the environment in which those principles are applied.

For B2B: Orchestration Across Complexity

In B2B, marketing unfolds across long, multi-stakeholder buying journeys that span teams, channels, and time. Performance depends on coordination across marketing, sales, product, and revenue functions. AIM OS acts as a force multiplier by providing shared intelligence and shared motion:

- *Adaptive Decisioning* identifies shifts from interest to intent and triggers timely outreach or sales handoffs.

- *Autonomous Activation* accelerates account-based and nurture programs, replacing calendar-driven drops with dynamic execution.

- *Experience Loop Intelligence* captures signals from content engagement, email, web activity, and sales interaction, feeding learning back into prioritization continuously.

- *Intelligence Governance* ensures workflows align with enterprise standards while making decision logic visible to stakeholders.

In B2B environments where silos often slow momentum, AIM OS becomes the connective layer that tightens alignment across revenue teams.

For B2C: Speed, Scale, and Signal Responsiveness

In B2C, interaction volume multiplies. Millions of customers, multiple product portfolios, and continuous engagement across channels. Performance hinges on how quickly signals are interpreted and relevance is delivered:

- *Adaptive Decisioning* determines next best actions in the moment, selecting offers, messages, and channels based on behavior and context.

- *Autonomous Activation* executes continuously across paid media, email, in-app, and web without manual coordination.

- *Experience Loop Intelligence* learns from every interaction, refining targeting, recommendations, pricing, and creative assets in motion.

- *Intelligence Governance* protects brand integrity and ensures privacy and personalization rules scale safely.

In B2C, AIM OS translates real time behavior into relevance across every interaction.

Exhibit 7: **AIM OS Across B2B and B2C**

Capability Zone	B2B Application	B2C Application
Adaptive Decisioning	Detect buying group intent and trigger timely sales engagement	Surface next best offer, message, or channel in real time
Autonomous Activation	Deploy account based and nurture programs based on live account activity	Launch and adapt campaigns dynamically across channels
Experience Loop Intelligence	Optimize outreach from engagement across content, email, and sales interaction	Continuously refine targeting, creative, and personalization
Intelligence Governance	Ensure workflows align with enterprise standards and visibility	Govern personalization rules, privacy, and brand standards at scale

Across both B2B and B2C, the underlying challenge is the same. AI now operates at every layer of the customer journey. Influence unfolds inside brand-controlled environments and increasingly inside AI mediated ones that summarize, compare, and guide decisions on the customer's behalf.

As influence spreads across both, complexity compounds in real time. Signals multiply. Context shifts faster. Decisions compress. Traditional marketing structures, built on team coordination, sequential handoffs, and periodic planning, cannot keep pace.

This isn't a failure of effort. It is a structural mismatch between how marketing is organized and how markets now move. AIM OS exists to close that gap by absorbing complexity into an operating model that senses, decides, activates, and learns faster than teams alone ever could.

For the CMO, this is a mandate to design how marketing works, instead of just what it delivers. For the CEO, it is a sponsorship commitment to support the operating model and the metrics that keep it healthy. Early results are already visible: shorter signal-to-revenue cycles, lower cost of acquisition with higher lifetime value, higher productivity without additional headcount, and fewer compliance surprises because guardrails are built in.

Summary

Across marketing, organizations are applying next generation AI within operating models designed for a different era. The mismatch is becoming harder to ignore. Leaders can see the inflection point, but many lack a disciplined path for crossing it. The answer isn't another platform, pilot, or isolated use case. It's an operating model designed for intelligence in motion.

The AI Marketing Operating System (AIM OS) provides that foundation. It integrates four interdependent capability zones into a single flow: Adaptive Decisioning, Autonomous Activation, Experience Loop Intelligence, and Intelligence Governance. Together, these capabilities define how modern marketing senses demand, makes decisions, executes at speed, learns continuously, and operates within clear boundaries. This is the structural shift underway as AI moves from a set of tools to the operating layer of marketing.

For CMOs, the implication is a redefinition of leadership. The role expands from coordinating activity to designing how marketing works. The objective isn't to automate effort, but to build a system that improves faster than the market changes. Creativity, strategy, and judgment are now even more essential, as they're amplified by an operating model that compounds precision and trust over time.

The choice facing leaders is increasingly binary. Continue running legacy marketing activity or design the operating system that defines modern performance. AIM OS represents that system, where intelligence is embedded by design and performance scales through structure.

The chapters ahead show how to build AIM OS, operate it at speed, and measure it so performance compounds.

Leadership Review

1. Are we still running marketing on effort and coordination, or redesigning it to move on its own?

2. Do leaders spend more time managing activity, or designing systems that continuously produce outcomes?

3. Where does latency still slow decisions, delay action, or drain talent?

4. When the next board review arrives, will we present more activity, or a marketing function with intelligence built in?

PART II

Building the AI Marketing

Operating System

5

The Rise of Marketing Engineering

The discipline that unifies data, martech, and AI investments

Marketing has always evolved. Engineering is simply the next layer. The future of marketing will be deeply technology centric. Agentic orchestration, data intelligence ecosystems, and generative AI are becoming the underpinnings of how modern marketing works. At the same time, marketing has always been repeatable. We create markets, build engagement, drive preference, and capture share. Those imperatives haven't changed. What has changed is the context in which they must be planned, executed, and delivered.

Today's marketplace is too complex and consequential to run on intuition and experience alone. Creative instincts still matter. Branding still matters. Differentiation still matters. Market influence still matters. But these drivers now operate within a larger frame of intelligent systems that cannot be manually sustained.

Modern marketing no longer competes on ideas, determination, or spend alone. It competes on the operating model that turns signals into decisions and decisions into action in real time. Creative, brand, and differentiation now run inside an intelligent system. Most organizations are still building toward this capability. We see AI experiments that remain scattered. Martech platforms stack. Data accumulates in silos. Marketing leaders face a pivotal choice. Either continue patching tools and teams for incremental gains, or build the discipline required to run marketing at the speed of the AI era.

Introducing Marketing Engineering (ME)

Marketing Engineering is the discipline that designs and runs marketing as an intelligent operating system. It connects strategy, data, and execution into one way of working, where data feeds intelligence, intelligence drives decisions, decisions allocate and activate, outcomes fuel learning, and governance keeps the system reliable.

Marketing Engineering is practiced by Marketing Engineers, the system architects of modern marketing. They design connected data layers, orchestrate AI models, integrate calibrated martech, and establish decision rights so the system responds in real time while protecting privacy and brand trust.

At its core, ME moves marketing beyond a fragmented set of tools and programs to a coherent architecture built for foresight, speed, and precision. Through integration, governance, and system design, it provides the foundation for marketing to compete and lead in an AI driven future.

In the sections ahead, we break down the four zones of the AI Marketing Operating System (Adaptive Decisioning, Autonomous Activation, Experience Loop Intelligence, and Intelligence Governance) to show how Marketing Engineering integrates and orchestrates these performance capabilities into one operating system.

Marketing Engineering is a distinct discipline from Marketing Operations. Operations manage the essential mechanics of the function, including financial management, procurement, planning cadence, resource allocation, and regional coordination.

*Marketing Engineering is a distinct discipline that blends
technology expertise with marketing experience.*

MEs focus is how workflows operate, how intelligence circulates, and how technologies are integrated. It doesn't replace marketing's mission. It amplifies it. It doesn't diminish creativity. It gives creativity scale. And it doesn't narrow the role of marketing leaders. It expands it by positioning them as system architects of growth.

Marketing Engineering makes the operating model real. It marks the line between tactical activity and structural advantage. For the first time in decades, marketing is reimagining not just what it delivers, but how it runs.

This discipline borrows its logic from engineering. Engineers design systems with purpose, versus a collection of parts. An aircraft isn't only metal and wiring. It's aerodynamics and propulsion. A robot isn't just a pile of sensors and actuators. It's embodied intelligence. In the same way, marketing in the AI era cannot be reduced to tools and tactics. It must be engineered as an intelligent operating system that can sense, interpret, decide, allocate, activate, and learn.

ME shifts the frame from tools to engineered capabilities. Many of the required technology layers are already familiar to CMOs. But in the AI era, they only create advantage when they work as one system. Data feeds models. Models guide orchestration. Orchestration drives activation. Analytics return feedback that improves the next move. Agentic orchestration brings cohesion to an ecosystem of specialized models, ensuring they act in concert rather than isolation. This is what Marketing Engineering is designed to deliver.

Exhibit 8: **How ME Delivers Engineered Performance**

Operating Layers	Modern Marketing Requirements
Data Layer	Customer and marketing data feeding decisions in real time.
Application Layer	Marketing and automation platforms connected as part of one system.
Orchestration Layer	Agentic and generative coordination across marketing-specific workflows.
Decisioning Layer	Agentic intelligence embedded into workflows to sense, reason, decide, allocate and act.
Governance Layer	Privacy compliance, brand standards, and model performance.

For CEOs, this reframes marketing from a cost center to an engineered growth system. For CMOs, it redefines and expands your mission to build the structure and system that uniquely aligns marketing with how AI reshapes customer decision making, and how marketing operates for competitive edge. The question for CMOs and CEOs is no longer "What tools do we have?" it's "What intelligent system are we building?"

The modern marketing question: Are your technologies still operating as individual support layers, or have they been engineered into a cohesive, intelligent operating system that learns and performs as one?

Why Is Marketing Engineering Emerging Now?

A CMO told me about a question he received from a board member during a coffee break. "I keep reading about brands accelerating their AI initiatives," the director said. "Generative creative production, predictive buying, confidence signaling. What's our play in all of this? Are you working with the CIO? Who's leading your team on these efforts?"

In that moment, it occurred to him that he couldn't point to a single person or team accountable for these questions. There were generative content tests and some predictive audience models, but they lived in pockets across his team. He couldn't name a single owner of AI on his team. And he couldn't speak well to how marketing was modernizing. He felt exposed.

That gets to the heart of the issue. Boards and CEOs aren't asking about tools or pilots. They are asking who owns AI marketing capabilities, how those capabilities scale, and how they drive resilient growth. Many CMOs don't have a clear answer yet. That hesitation is understandable. AI represents a profound shift. But boards and CEOs already recognize AI as a disruptive force, and they expect to see the same urgency reflected in marketing leadership.

This is why Marketing Engineering is
emerging as a structural necessity.

The context CMOs operate in has crossed a threshold. Complexity, speed, and technology have converged to a point where marketing can no longer rely on siloed tools or a loose community of technologists. Modern CMOs don't view technology as a supporting lever. They view marketing technology as a strategic imperative.

If technology isn't changing how you run marketing, then you have a collection of tools, versus an engineered system. Five forces now make the need for Marketing Engineering inevitable. These are:

1. *Martech sprawl:* The average enterprise uses more than 65 marketing tools. They sit in silos, stitched together with manual workarounds. The result isn't a system capability. Marketing Engineering brings cohesion, aligning those investments, optimizing their capabilities, and aligning them into a system that performs better than the sum of its parts (more on efficiency in Chapter 12).

2. *AI and autonomous readiness*: AI now drives real time decisions,

generates content variations, and optimizes engagement at scale. It's the new operating layer of marketing. A structured process to deploy, train, and govern agentic reasoning and decisioning capabilities turns AI experiments into scalable impact. Marketing Engineering is the discipline that makes AI productive and governable (more on this in Chapter 9).

3. *Data explosion*: Data is growing at an unprecedented scale. Left unmanaged, it becomes fragmented, inaccessible, and slow to yield value. Marketing Engineering builds the system to activate structured and unstructured data, creating the foundation where analytics deepen and AI turns this volume into advantage (more on data in Chapter 11).

4. *CFO driven accountability*: Marketing is under sharper financial scrutiny than ever. Boards and CFOs expect line of sight from spend to business outcomes. Marketing Engineering delivers this by embedding measurement and telemetry into the AIM OS system, so performance isn't argued after the fact, it's visible as it happens (more on this in Chapter 10).

5. *Customer expectation shifts*: Customers expect relevance, immediacy, and seamless experiences. They care about experiences that adapt to their needs in AI time. Marketing Engineering makes this possible by aligning intelligence, content, and orchestration.

Taken together, these forces create a point of no return. The layered ways marketing has historically been managed can no longer keep pace. The future of marketing runs on technology, and marketing's reach travels only as far as the operating structure that delivers them.

That is why Marketing Engineering is rising now. It turns disjointed technology into strategic capability, converts AI experiments into performance systems, and translates signals into competitive advantage. The Marketing Engineer isn't optional. It's the role that makes modern marketing possible.

The Marketing Engineering Mission

Marketing Engineers are hybrid leaders fluent in marketing strategy and execution, and equally fluent in AI, data, martech, and automation. Their role is to remove the friction that prevents marketing from operating with speed and scale. Stacks may look impressive on a slide, but without integration and intent, they create complexity rather than capability.

Marketing Engineers accelerate how marketing gets done in the AI era.

A collection of instruments becomes music only when they are tuned, arranged, and conducted to the same score. Many martech investments today remain optimized around department and discipline level silos, which limits AI's potential rather than unlocking it. Marketing Engineers provide the orchestration that tunes, arranges, and conducts the AIM Operating System. The question is simple. Do you have a collection of tools, or an orchestrated system?

The Marketing Engineers provide two layers of immediate advantage:

ME Mission 1: Get Fit

Most marketing teams believe they already have the technologies they need. The platforms are in place. The licenses are paid for. The dashboards exist somewhere. Yet when performance stalls, the reflex is often to add a platform to the pile. Another "must have" point solution, with vendors promising a breakthrough. The result is what many leaders now quietly recognize, that a random martech stack isn't a technology strategy.

I met a new CMO who was shocked to discover her new team didn't even know how many platform licenses they held or who was managing them. I knew her from a previous assignment, and she was very tech savvy. A quick audit revealed dozens of tools with no oversight, some barely used at all. What looked like a stable foundation was actually and ungoverned collection of platforms.

It's important to note that, this isn't a critique of martech platform offerings. In fact, quite the opposite. Virtually all the martech partners we work with are frustrated by the constant churn that comes from underutilized capabilities. They want to be essential to their clients. They don't want to be relegated to fringe usage. Because to them, fringe means the risk of lost value, higher cost, and the dreaded churn. "Please help our clients use this platform better," they tell us. Partners want utilization because underuse is a value and retention risk for them too. Tools without purpose and orchestration don't create capability. They create clutter. And the more fragmented the stack, the more effort goes into patching, reconciling, and explaining rather than producing impact.

Efficiency is the foundation for modernization.

Before teams can modernize, they must first stabilize and streamline their existing operations. Fit systems reduce waste, save valuable resources, and create the runway for innovation. What does "Getting Fit" look like? Four core capabilities define it:

1. *Systems Thinking*: Marketing Engineers look across martech tools to orchestrate and architect work end-to-end. They design how data, content, and decisioning flow across teams, with clear decision rights and standards.

2. *Data & Intelligence*: AI cannot perform without clean, connected, and usable data. MEs ensure data integrity, accessibility, and usability. They design pipelines and feedback systems that turn signals into decisions.

3. *Automation & AI Integration*: Marketing Engineers augment repetitive tasks and place them in automated flows. They embed agentic intelligence into execution so marketing can adapt in real time, and self-improve.

4. *Cross Functional Orchestration*: No marketing system stands alone. Marketing Engineers collaborate closely with vendors, IT and Operations to ensure the operating system stays in-sync with other enterprise teams.

When these foundations are in place, the effect is immediate. Waste declines and visibility improves. Your marketing team moves with less friction. But more importantly, marketing leaders gain the runway to modernize. We'll explore this dynamic even further in Chapter 10.

For now, remember that efficiency is more than a margin contribution exercise, it's the platform that makes innovation possible. Without it, AI remains stuck in pilots, orchestration remains a theory, and marketing's role remains reactive. With it, the path to modernization and building the AI Marketing Operating System opens.

ME Mission 2: Get Modern

Once the clutter is cleared, workflows stabilized, and data usable, the real opportunity emerges. This is where marketing stops accumulating and starts engineering for growth and competitive edge.

Modernization is orchestrating what your already own into an intelligent system with purpose. It's the practical convergence of data pipelines, martech platforms, and AI orchestration into a unified way of working. This is the foundation of your AI Marketing Operating System (AIM OS), shifting marketing from activity to architecture, and from scattered tools to engineered capabilities. Everyone wants to jump right in, but if you skip the "Get Fit," stage, progress stalls and your initial AI moves will multiply inefficiency.

AIM OS is a cohesive system that integrates data, martech, and AI into the four performance zones of capability: Adaptive Decisioning, Autonomous Activation, Experience Loop Intelligence, and Intelligence Governance.

Getting modern is about putting your technology to work with a clear purpose of growth that is engineered, orchestrated, and adaptive. The AIM OS zones reinforce each other, creating a system that learns, adapts, and compounds performance. Intent shapes experiences, experiences feed activation, activation generates new signals, and governance ties it all together. The system will create faster signal-to-action time, higher decision-automation rate, higher creation/reuse of assets, and lower unit cost of testing. If these topline KPIs across your function don't improve, then your operating model is dated. The mission is marketing powered by engineered performance, and not just effort.

For CMOs, modernization marks the shift from proving value with isolated pilots to proving impact with systems that scale. For CEOs, it means marketing is no longer a cost to explain but a designed operating system for growth.

Engineering AIM OS Through Orchestration

Now let's get pragmatic. Breakthrough performance in modern marketing comes from how existing and emerging technologies are integrated and orchestrated into one intelligent operating system that can learn in motion.

When three foundational layers converge (martech platforms, marketing data, and artificial intelligence) they unlock system-level capabilities that now define competitive performance. Each layer is essential on its own. But only when engineered together, with intent and governance, do they become the AI Marketing Operating System. Most organizations already own the components. What's missing is the orchestration and integration expertise to align them to business strategy and customer behavior.

Exhibit 9: **The Engineered Intersections
Forming AIM OS Capabilities**

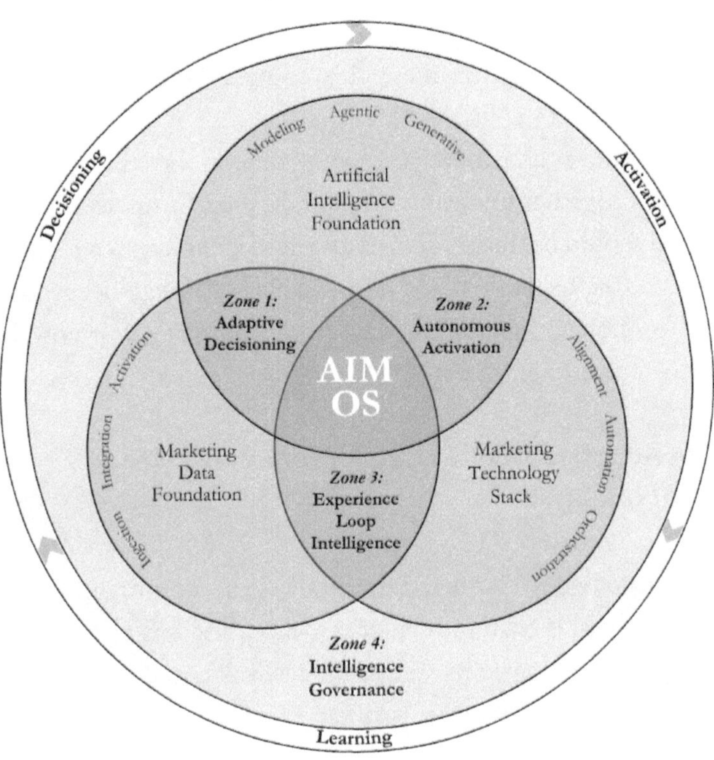

Bridging these gaps requires new capability inside marketing. Traditional roles, even with strong platform knowledge, weren't designed to build or maintain connected systems that operate in real time. Strategy, brand, program management, digital, content, and operations teams are rarely equipped to design adaptive workflows, deploy agentic AI, or engineer data driven reasoning and decisioning.

That is why activating AIM OS requires Marketing Engineers. They connect platforms, data models, and marketing logic into cohesive systems built for speed, scale, and learning. These systems are formed at engineered intersections where signals meet models, models drive workflows, workflows activate channels, and data meets governance. This is where modern marketing capability emerges.

Getting modern means building this operating system deliberately.

The mission of Marketing Engineering is to help the function get fit, then get modern. Efficiency first. Intelligence next. Order, then acceleration. In an environment this complex and this fast, growth isn't improvised. It's engineered.

In the sections ahead, we go zone by zone to show how these capabilities work together to create compounding performance. Each capability zone should be understood as an engineered intersection forming from marketing technologies. And in Chapter 9, we go deeper into the specific technology foundations that power each one. For now, here is how each zone is engineered:

Zone 1: Adaptive Decisioning = *Marketing Data + Artificial Intelligence*

Adaptive Intelligence is the operating system's command center, the layer where signals turn into choices. Marketing has always relied on strategic intent, key insights, consumer feedback, and market research. This valuable data, however, is often spread across departments or shared in planning meetings. In the AI era, that input is engineered into the system for a more real time, data driven, adaptive approach.

For Marketing Engineers, the work starts with *signal and intent architecture*. Data exists everywhere, like CDP and CRM, web analytics, call centers, primary research, campaigns, satisfaction surveys, and partner data. It's scattered, inconsistent, and often unstructured. Marketing Engineers design pipelines to clean, align, and stream this data into unified data layers. Without this data intelligence groundwork, "AI" is just algorithms sitting on top of random data sets.

Second is *model integration*. MEs operationalize models, whether built in house or licensed. They translate output into workflow where a churn score becomes a retention trigger, a propensity score becomes an offer, a content model becomes a recommendation.

Finally comes *orchestration logic.* This is where intelligence becomes action. MEs design rules and guardrails with decision rights and standards that specify what happens next when a signal triggers a model. Does the system allocate budget, send a message, shift spend, or escalate to management review. Done right, this creates a feedback system where every action sharpens the next decision.

For CMOs, this means moving from decision meetings to decision systems. For CEOs, it means faster adaptation and reduced waste with new metrics like signal to action time and decision automation rate. For Marketing Engineers, building an intelligent decision fabric that analyzes and acts under governance is essential.

Zone 2: Autonomous Activation = Artificial Intelligence + Martech Platforms

Marketing is execution heavy with a tremendous number of messages, offerings, creative assets, and channels that require a lot of coordination. Autonomous Activation changes that by turning signals into actions in real time. Instead of manually developing and implementing tactics, systems read signals and autonomously apply tactical responses.

Marketing Engineers deliver this by designing *activation frameworks* with decision rights and standards. They connect content systems, ad platforms, and workflow tools so outputs aren't isolated but coordinated. The system knows which assets exist, where they can be mobilized, and under what conditions.

Next, they embed *automation and AI* into the operating system. MEs configure rules and models so campaigns no longer rely on manual scheduling or one-off pushes. A model predicts interest, the system launches an ad, and feedback adjusts spend in motion.

Finally, engineers ensure *governed autonomy.* Not everything should run unchecked. MEs design sample inspections (human-in-the-loop), overrides, escalation paths, and monitoring so oversight stays in the flow where it matters most. Guardrails are brand safe, privacy compliant, and outcome aligned.

For CMOs, this means shifting team resources from manual campaign execution, to designing strategy and systems. For CEOs, it means marketing scales without always scaling headcount. For Marketing Engineers, it's creating an activation layer that runs continuously.

Zone 3: Experience Loop Intelligence = *Martech Platforms + Marketing Data*

Consumers never see all of the mechanics behind the curtain. They experience the brand. The test is whether every touchpoint feels timely and relevant. Experience Loop Intelligence is the operating system's listening and memory layer. It learns from interactions and feeds that learning back in real time.

Marketing Engineers make this happen by building *feedback architectures*. Every click, view, download, or purchase must feed reporting, experience design, and decisioning. This includes standards for event taxonomy, identity, and consent.

Next, MEs design *experience models*. This ensures creative, offers, and journeys adapt based on signals and intent. MEs connect content systems, CDPs, and orchestration platforms into flows where content adjusts dynamically under clear guardrails. Models align to business and program objectives.

Finally, they close the *feedback system*. Data flows back out to refine. Engineers ensure insights feed models, update decision logic, and shape the next activation.

For CMOs, this shifts CX from static programs to dynamic feedback systems. For CEOs, it builds stickiness and lifetime value. For Marketing Engineers, it's designing living systems that get smarter with every interaction.

Zone 4: Intelligence Governance

Without governance, intelligence turns into risk. Black box models, biased data, and unmonitored automation can erode trust faster than they create value. Intelligence Governance ensures systems are explainable, auditable, privacy compliant, brand safe, and aligned with outcomes through clear decision rights and standards. Said another way, is your AIM OS reasoning and making the correct decisions?

Marketing Engineers lead this by designing *data and model oversight*. They set standards for data quality, lineage, identity and consent, bias checks, drift monitoring, and model registries. They ensure models can be interrogated so it's understood why a decision happened, whether it's repeatable, and what data was used.

Second, they create *system transparency*. MEs build dashboards not just for performance, but also for operations. This includes what signals are flowing, what actions are triggering, and what models are running. This visibility is critical for leadership teams and boards.

Finally, MEs embed *guardrails and controls*. They design logic that caps frequency and ensures compliance with privacy and brand standards. This includes kill switches, rate limits, and human-in-the-loop thresholds where required. Proper governance is what makes autonomy safe.

For CMOs, governance builds confidence that AI won't run off brand or off the rails. For CEOs, it creates resilience and accountability. For Marketing Engineers, it turns fragile experiments into trusted systems.

Together, these four zones move marketing from experimentation to engineered performance and make the role of the Marketing Engineer indispensable.

Marketing Engineering as Competitive Edge

At its core, Marketing Engineering is a growth discipline. It defines the conditions under which performance becomes resilient. Operations manage activity. Engineering designs the architecture that makes success repeatable and measurable.

Think of it this way. Campaigns come and go. Channels shift. Tools evolve. But the system gets smarter and more productive. The functions that thrive tomorrow won't be remembered for their clever campaigns. They will be defined by the AIM OS they built, and the compounding performance they deliver. This reframing matters deeply for leaders.

Marketing Engineering must be viewed as strategy, versus support.

For CMOs, it shifts the mandate from producing marketing to designing the conditions in which marketing performs. Your role becomes more than a marketing leader, but now also system architect leader too. For CEOs, it reframes marketing from a discretionary spend to a designed operating system for growth, an asset that compounds, versus a cost that fluctuates.

Summary

Marketing has entered an era where performance is increasingly driven by the quality of the operating system that connects intelligence to action. As AI reshapes how customers decide and how markets move, marketing must operate as an engineered system rather than a loose collection of tools and campaigns. It must be engineered. The AI Marketing Operating System is engineered by integrating and orchestrating foundational technology functions already owned, including data, martech platforms, automation, and AI agents into modern performance capabilities. Most enterprises have these components. What's required is the system design and integration to make them work as one. This is the work of Marketing Engineers.

*Marketing Engineering is the discipline that designs
and runs marketing as an intelligent operating system.*

The work has two deliberate missions. First is get fit. This stabilizes the environment by reducing sprawl, integrating fragmented data, clarifying ownership, and removing friction. Second is get modern. This is where AIM OS is engineered as a unified way of working, with intelligence embedded directly into how marketing operates every day.

Through AIM OS, Marketing Engineering brings together foundational technologies at critical intersections to form the four performance capability zones (Adaptive Decisioning, Autonomous Activation, Experience Loop Intelligence, and Intelligence Governance). These capabilities are emergent. Performance improves as intelligence flows across zones and compounds across the system. Decisions move faster. Activation becomes adaptive. Risk is governed in flow. Marketing begins to operate as a coordinated system rather than a set of disconnected functions.

Engineering is how marketing now holds together. It becomes repeatable and scalable. For CEOs, marketing becomes an engineered growth system. For CMOs, the role expands from steward of activity to architect of the operating system that powers performance.

The defining question now is who will own and lead the system that makes modern marketing work. The answer points squarely to the CMO.

Leadership Review

1. Who on the leadership team owns the operating system that makes marketing work as an intelligent system?

2. Do we have a named Marketing Engineering leader with authority, scope, and budget tied to outcomes?

3. Have we defined where Marketing Engineers will come from, and how they will be organized?

4. Do we have an AIM OS roadmap that aligns existing capabilities, integrates intelligence, and tracks system maturity?

5. How will we ensure AI aligns with our business priorities, strengthens our brand equity, and accelerates our marketing impact?

From Tools to Engineered Capabilities: A Diagnostic Framework

Most marketing teams aren't short on technology, but they're short on leveraging these investments with orchestrated strategic intent. To move from fragmented execution to engineered performance, leaders must start asking better questions: From: *"What platforms do we own?"* To: *"Where are the system-level gaps preventing us from creating strategic capability?"*

The table below outlines how the four performance capabilities emerge from specific system intersections and what to examine when diagnosing weak points.

Exhibit 10: **Performance Capability Zone**
(Intersection & Diagnostic)

Capability Zone	Key Technology Intersections	Diagnostic Questions
Adaptive Decisioning	CDP + AI models + Real time behavioral signals	Can we process and respond to customer behavior as it happens?
Autonomous Activation	Workflow automation + Dynamic content + Channel orchestration	Do campaigns deploy automatically based on signals, or still manually executed?
Experience Loop Intelligence	Analytics + Event tracking + Feedback into model layer	Are we learning and adjusting while campaigns run, or only after they end?
Intelligence Governance	Data observability + Model explainability + Compliance logic	Can we see how decisions are made, and is system aligned to brand, privacy and policy standards?

When any one of these intersections is weak, performance slows. When two or more are missing, the system cannot sustain. That's why advantage doesn't come from marginal upgrades or tactical investments. It comes from engineering the operating system, connecting these intersections so that intelligence flows through them continuously.

6

The CMO Role as System Leader

Leading the system architecture shift to modern marketing

The CMO stands at a crossroads. On one side lie the familiar marketing activities like planning, positioning, and coordinating. On the other lies something less familiar but far more decisive. This is the mandate to architect the operating system that will define performance in the AI era.

For years, CMOs have managed the brand, remained customer focused, supported business units, championed creative insight, driven revenue growth, and embraced new technologies. Each responsibility stretched the role wider. Now expectations have shifted again. Boards and CEOs are asking whether AI is being engineered into marketing to create a measurable advantage. They're also looking for assurance that the company is keeping pace. Here are some of the questions we're hearing: How are we using AI to create differentiation in the market? What are competitors doing that we are not? How does this tie directly to revenue growth? What safeguards are in place to ensure brand safety and ethical use of AI? Do we have the skills and structures in place to use AI effectively? What training or new roles are required to make AI part of how we operate?

In essence, they are pushing marketing leadership to think beyond traditional marketing toward organizational capability and culture. Boards aren't satisfied with hearing "We're doing AI pilots." They see the flow of AI capital investments. They understand valuation models. They worry about being disrupted. Many have personally managed through previous inflection points. They know what's at stake, and they know what style of leadership is required to succeed in this new era.

Marketing now competes on the capability of its AI Marketing Operating System. Creative, brand, and differentiation remain essential marketing factors, but they now run inside an intelligent system. This is why marketing leaders face a pivotal decision. Either keep patching tools and teams for incremental productivity, or build the discipline that operates at artificial intelligence speed and scale. As legendary marketing author and consultant Philip Kotler said, "There will be two kinds of companies: those who change and those who disappear." Today, AI is the structural change that determines which side you're on. This is the shift, and it's now the CMO mandate. Marketing leaders who hesitate give up competitive ground while competitors gain advantage.

Yet many CMOs feel unprepared for this moment. They are brand builders, strategists, and operators. They sit next to CIOs and CTOs who are fluent in enterprise applications, cloud architecture, data management, and emerging technologies, and wonder whether the mandate aligns there instead.

The hesitation is understandable, but it's also misplaced. CIOs and CTOs are critical partners, but they cannot deliver customer growth. They can enable, but they don't own marketing's mission or functional strategy. That responsibility belongs to marketing, and specifically to the CMO. You're the only executive who understands the scope and dynamics of marketing including what the marketplace values, what motivates consumers, and how ideas become programs.

CMOs are the domain experts who translate business priorities into customer impact. Without that translation, no amount of technology expertise will matter.

While CMOs aren't expected to be system engineers, they are expected to be system architect leaders. The good news is every organization already sits on the raw ingredients of an AIM OS. What's required is leadership that can bring those ingredients together into an orchestrated, intelligent system. This is a structural decision. Performance doesn't come from what you have, it comes from *how* it's designed to work together. This chapter is about what it means to step into that role.

The Weight of Expectation

The mandate increasingly surfaces every time a CMO sits in front of the CEO or Board. The questions are shifting quickly. In the past, the focus centered on activity. How are we executing against this strategy? What campaigns are in market? How much pipeline was created? How's the brand tracking? Today, the questions are structural. How are our competitors using AI? Can we use AI to create a competitive edge? Why are we hiring more marketing staff? Can we scale marketing without scaling cost? How is marketing fundamentally changing because of AI?

This shift reframes the role of the CMO. Activity alone no longer sustains advantage. Programs may generate consideration spikes, but aren't easily sustainable. A system, by contrast, learns with every interaction. It senses, interprets, decides, allocates, activates, and learns by turning signals into intelligence, intelligence into action, and action into advantage. The rising expectation from the top is that marketing moves from campaigns to continuous, compounding performance.

The competitive cost of delaying this marketing shift is very real.

Every quarter spent tinkering with pilots and tools rather than engineering the AIM OS is a quarter in which competitors expand data intelligence ecosystems, raise decision automation rates, extend autonomous activation, and build new marketing engineering skills. We interviewed an executive who had recently joined a former competitor. What surprised him most was not strategy or talent, but how far behind his new organization was in system learning. Capabilities that he assumed were just table stakes were not happening at all in how the company used AI to learn and improve. He questioned if his new team could actually ever catch up. These capabilities define modern performance. And learning is where the gap widens.

CEOs want to see marketing embrace AI at the core and run it as a system that is intelligent, scalable, brand safe, and directly accountable for growth. This is the weight now resting on the shoulders of every CMO. And any delay transfers advantage to competitors who build essential skills and consistently fine-tune their operating system.

A Tale of Two AI CMOs

Recently, I was advising a leading executive recruiter who was searching for a new CMO. The job description was ambitious. The role required AI experience, with demonstrated operational proof that the leader could move the company's marketing function into the future.

The recruiter asked me a simple but loaded question: "How do we know which candidates truly understand AI, and which ones are simply adding it to their LinkedIn profile?" It was a conversation I've had more than once. CEOs are asking the same thing. The market is flooded with claims of "AI experience," but the depth and meaning behind those claims vary widely.

I told the recruiter there are two kinds of leaders who describe themselves as AI skilled CMOs. The first group are *Productivity CMOs*. They use AI as a booster inside the old model. They're still running marketing in traditional ways, and they're layering generative tools to write faster, design quicker, or automate routine tasks. Some are using agentic AI features inside existing martech platforms for automation or CRM. All are very valuable. They make existing activity more efficient. But the operating model doesn't change. Adopting tools, yes. Shared team commitments, no. Structure change, no.

The second group are *System Leader CMOs*. They reinvent workflows with marketing talent *and* marketing agents. They're building AIM OS and recruiting marketing engineering talent as a discipline. They're training teams to work differently, embedding AI into decisioning, activation, and governance. They've created a culture where AI isn't a tool on the side, it's the operating layer of marketing. These leaders talk about reinvented roles, team alignment, iterative progress, and step changes in AI enabled performance. Adopting capabilities, yes. Shared team commitments, yes. Structure change, yes.

That distinction clicked immediately with the recruiter. We walked through a sample of candidates and reviewed how they described their work. Did they talk about AI in terms of incremental productivity and efficiency, or in terms of shifts to intelligence, decisioning, and performance? Did they generally speak about AI as an add on, or as a structural foundation? Could they name who owned the operating system and the marketing engineering talent on their team? Those conversations quickly reshaped the shortlist.

For Boards and CEOs, this is a critical lens. The future of marketing will be defined by leaders who reimagine the operating model, align the team, and accumulate the skills needed to engineer the system. It won't happen overnight, but the markers are very clear, and they indicate whether CMOs and their teams are on the right trajectory.

CMOs facing the AI mandate tend to fall into two groups. Some see AI as a way to do more with less. Others see it as the chance to re-architect how marketing runs. The distinction matters because one produces busier teams, and the other builds enduring advantage. What side are you on?

Exhibit 11: **The Two Types of AI Focused CMOs**

Dimension	AI Productivity CMO	AI System Architect CMO
Focus	Uses AI to cut costs, automate tasks, and increase output.	Engineers AI into the OS to compound performance and build competitive advantage.
Approach	Deploy pilots and platform features across isolated teams.	Orchestrates integrated systems across the four performance zones.
Evidence	Activity gains (e.g., more content, faster campaigns).	Structural gains (e.g., shorter cycle time, compounded ROI).
Metrics	Launch and volume (e.g., launches, assets, reach).	System metrics (e.g., signal-to-action time, decision automation rate, reuse rate).
Organization	Keeps traditional roles, structure, and vendors.	Team aligned to AIM OS, Marketing Engineers partner with functional marketing.

How Marketing Engineering Makes This Possible

If this mandate feels new, it is. For decades, marketing was managed as a collection of disciplines like brand, programs, products, verticals, segments, channels, and experiences, each operating on its own rhythm and reporting its own status and metrics. Integration remained more aspiration than reality. Marketing running as a single, intelligent operating system stayed out of reach.

Traditional marketing was deeply customer focused. It was built on strong customer understanding through research, segmentation, brand strategy, and experience design. The limitation was often scale. Those insights lived in plans, presentations, and campaigns rather than in systems that could sense behavior, decide, and adapt in real time.

What has changed is the rise of Marketing Engineering. It encompasses the marketing application of advanced data intelligence engineering, agentic workflow automation, AI decision chains, and autonomous activation. Customer signals now flow across platforms in real time, are interpreted by models, drive decisions, allocate resources, activate execution, and feed learning.

This work is practical and hands on. It includes cleaning and normalizing data, mapping sources and identity, and fixing quality issues that erode trust. It includes managing martech platforms and calibrating, configuring, and repairing systems that fall out of alignment with goals. It includes setting up and governing AI agents, training them with the right logic, monitoring for drift and model registries, and ensuring recommendations align with brand and business strategy. It includes building workflows that automate and improve continuously through testing, tuning, and iteration.

The scope is significant. Consider manufacturing without Industrial Engineers, chip design without Electrical Engineers, or medical discovery without Biomedical Engineers. The strategic potential of AI, paired with the discipline of Marketing Engineering, creates the conditions for marketing to operate as an engineered operating system. This is why the mandate now belongs to the CMO.

You Don't Need to Replace Your Team

Many CMOs worry that their teams lack sufficient technical depth or that they don't yet have the right AI skills. These concerns are understandable, but they're misplaced. Marketing Engineering exists to unlock more of the team's potential.

*Your marketing teams hold the domain
expertise, and that matters most.*

They know your customers. They understand your brand. They've built the programs, content, and channels that keep the business in motion. None of that disappears. The team's expertise becomes more valuable when amplified by intelligent systems. What changes is the operating context. Instead of manually pushing programs, teams collaborate with Marketing Engineers to embed their knowledge directly into the marketing operating system.

Your team's judgment becomes the logic that informs decisioning models. Creative instincts fuel systems that activate dynamically. Understanding of customer journeys shapes experience feedback systems that improve continuously.

Hybrid roles are already emerging across modern teams. A content strategist who once debated copy variations now guides a generative system capable of producing thousands of branded messages. A campaign manager who once worked from static calendars now oversees a portfolio of self-optimizing journeys. A journey designer who once mapped flows on whiteboards now collaborates with workflow engineers to automate and adjust paths in real time. These are role amplifications. The system handles repetition while the team provides strategy and judgment.

*Your functional expertise remains essential.
What changes is how it's executed.*

This is why the most successful CMOs rebuild their organizations rather than tear them down. They make space for hybrid roles where marketing acumen and technical fluency meet. They pair creative strategists with data modelers, journey designers with workflow engineers, and ensure system ownership stays with marketing even as new capabilities are introduced. CIOs and CTOs continue to enable and support. Ownership of this shift rests with the CMO.

From a CEO's perspective, this approach matters. Rip-and-replace strategies create disruption, knowledge loss, and demoralized teams. Elevating and evolving the team protects institutional memory, retains customer expertise, and compounds value. This approach gives the organization continuity while building capability. AI transformation succeeds when it compounds productivity and performance on top of the existing foundation, not when it pursues job replacement. Leaders who shift resources blindly toward AI risk losing business understanding, decision logic, and creative instincts that are essential for training, building, and scaling the AIM OS.

Elevate and evolve your team with customer intelligence and brand expertise that no outside vendor or LLM can replicate. What they need is CMO conviction and Marketing Engineering partners who share the same language and commitments.

This reminds me of a leadership quote from Duke basketball legend, Coach Mike Krzyzewski, *"I get a group of people who are talented to commit to excellence and to work together as one. That's where it starts. Different talents, same commitment."* Leadership in the AI era aligns proven marketing expertise, marketing engineers, and AI agents so marketing performs as one committed modern marketing team.

Leading at The Inflection Point

Today, the modern CMO carries two mandates. The first is to run marketing now, including brand consideration, customer engagement, pipeline, and market share growth. These are the business results your organization expects quarter after quarter, and where credibility is earned. The second mandate is to build the system for what's rapidly coming next: a modern structure and the AI Marketing Operating System that will redefine how performance is created.

These two mandates don't run in sequence. They run in parallel.

This responsibility is demanding and essential for protecting your brand's market position. Growth targets can't pause while the system is built, and system design can't wait until after the quarter closes. Credibility comes from delivering today and protecting the advantage created by building AIM OS. When either mandate is neglected, performance suffers.

Other executives have navigated the same duality. CIOs kept legacy systems running while re-platforming enterprises to the cloud. Many recall patching servers to keep operations steady while defending million-dollar migrations. CFOs closed the books every quarter while modernizing finance into a real time function. COOs maintained performance while redesigning operations for leaner, more automated execution. Manufacturers sustained production while industrial engineers rebuilt systems around them.

For CMOs, the dual mandate is executing marketing to meet current business goals while adapting to the AI era. Each day the system remains unmodernized creates space for competitors to improve data intelligence, raise decision automation rates, test autonomous activation, and compound advantage. CEOs are beginning to ask two questions: "What did marketing deliver this quarter?" and "How ready is the AI operating system to deliver more next quarter?"

The CMO agenda therefore operates on two cadences. The first is tied to immediate results. The second is tied to system maturity metrics such as signal-to-action time, decision automation rate, reuse rate, and data quality. Progress against both mandates must remain visible to leadership teams.

CMOs who manage the dual mandate well establish two languages of accountability. The first is tactical and near term, measured through familiar metrics like sales contribution, brand equity, and engagement. The second is structural, measured through system maturity indicators such as build progress, team alignment, operating metrics, telemetry, and data quality. Keeping both in play gives CEOs and boards confidence that marketing delivers results today while building capability for tomorrow. We'll explore these metrics further in Chapter 10.

The weight of this dual mandate is significant, and it defines your opportunity. Treating AI as the foundation while sustaining current performance places CMOs squarely in the spotlight. Few roles in the C-suite now carry greater visibility or influence in shaping customer decisions and the trajectory of that growth.

The CMO as System Architect Owner

Owning the system elevates the CMO from functional leader to system architect. Marketing now owns the operating system for growth (AIM OS) that connects the market with sales, product, service, and finance through a common flow of data, signals, decisions, allocation, activation, and learning. No other executive operates at this intersection. This is why the role belongs within marketing rather than being delegated to IT, a software vendor, or an agency.

Only the CMO ensures the system is designed to deliver the organization's vision, culture, and business priorities.

And when this role performs well, the impact cascades across the enterprise. Sales teams benefit from cleaner signals and sharper prioritization, improving pipeline conversion. Product teams gain clearer feedback from market interactions, shortening innovation cycles. Service teams operate with richer context across customer journeys, strengthening retention and upsell. Finance gains predictability as marketing evolves from discretionary spend to a measurable, system driven investment with line of sight from spend to outcomes. In every case, operating system ownership in marketing strengthens enterprise performance.

The CMO who embraces this role sets the tempo of growth for the entire business. As the AIM OS owner, the CMO defines which signals matter, how content activates, how workflows connect, and how intelligence is governed. In partnership with the CEO, they protect the transformation from short term pressures and compromise. Instead, they move AI beyond isolated pilots and into the operating fabric of modern marketing.

A CMO acting as system owner shapes the growth architecture of the enterprise. They build the operating model that determines whether growth compounds or stalls. CEOs gain more from asking how system design is advancing, whether the architecture is ready for the next phase of scale, what resources are required, versus from focusing only on new campaigns or short term sales contribution.

The AI Marketing Mindset

To step into system ownership, CMOs must lead with a new AI marketing mindset that shifts the culture of marketing. Here are some emerging attributes of a modern marketing culture:

- *From output to outcomes:* Success is measured not in the volume of campaigns delivered, but in the performance of faster cycle times, higher conversion rates, and lower cost per acquisition.

- *From intuition to signals:* Decisions are grounded in real time feedback systems and model learning, not just creative instinct. Judgment remains essential, but it is informed by continuous intelligence.

- *From linear to continuous flows:* Workflows are designed to improve continuously, not reset campaign by campaign. The system gets smarter with every interaction.

- *From tools to architecture:* Martech is no longer a stack of disconnected platforms. It becomes an integrated architecture with decision rights, standards, and telemetry, engineered to create compounding advantage.

- *From individual wins to system health:* The measure of success shifts from whether one campaign "worked" to whether the system accelerates.

With this mindset, the AIM OS becomes the foundation of modern marketing, that senses, interprets, decides, allocates, activates, and learns, amplifying every skill, every decision, and every investment.

This Is a Leadership Challenge, Not a Technical One

CMOs often hesitate because they assume building an operating system for marketing (AIM OS) requires expertise they don't yet possess. The essential work centers on setting the vision, building conviction for change, and aligning culture to a new way of working.

The obstacle to modern marketing isn't technical literacy, it's CMO conviction.

Technical execution sits with Marketing Engineering specialists such as data engineers, AI architects, and martech specialists. The CMO owns the vision and architecture, and they keep progress visible at the executive table. This is a management challenge before it becomes a technical one.

The most persistent barriers are vision and incentives. Legacy routines slow workflows that should move in real time. Incentives tied to campaign volume reward activity over impact. Traditional reporting reinforces silos rather than surfacing orchestrated system performance.

Which is why this is first and foremost a leadership challenge. The hardest part isn't acquiring technologies, because anyone can buy them. The hardest part is leading people through change. It's the ability to set a compelling narrative of why marketing must operate differently, and to hold that course when short term pressures may prompt a default to traditional ways of working.

For CEOs, this distinction matters. CMOs should be measured on their ability to lead marketing into the AI era, not on technical credentials. The core skill is leadership and culture, the kind that converts AI vision into discipline and discipline into results.

The most dangerous position for a CMO sits in the middle. Delivering against today's priorities while drifting into partial modernization. Boards see motion without momentum. CEOs hear the promise of AI without seeing it reflected in results. Teams feel the churn of initiatives without the acceleration of change. The outcome is fragmented adoption that creates the appearance of progress without delivering system performance.

For CMOs, the career stakes match the business stakes. This is the leadership test that defines trajectories in the AI era. Those who step into AIM OS ownership transform marketing and position themselves as enterprise leaders, with potential paths toward CEO succession. And those who hesitate risk being remembered as the last generation of traditional marketers rather than the first generation of AI marketing builders.

Summary

The defining question facing CMOs is whether marketing leadership will own the system that now determines performance. AI is already reshaping how customers decide and how markets move. What matters next is who will take responsibility for engineering that advantage.

Marketing has crossed a structural line. Boards and CEOs are no longer seeking reassurance that marketing is experimenting with AI. They are looking for evidence that marketing is being run as a coherent, intelligent system that can scale, adapt, and remain brand safe under pressure. They want confidence that growth is built on an operating model designed for resilience.

Some CMOs will treat this moment as a productivity upgrade, using AI to accelerate existing workflows and reduce cost. Others will recognize the deeper shift and step into system ownership, redesigning how marketing operates by engineering the AI Marketing Operating System. The difference is not technical skill. It is leadership intent. One preserves the past. The other builds the future.

Marketing Engineering makes this transition work. It provides the discipline that turns customer understanding into system logic, connects data and platforms into unified workflows, and embeds governance so intelligence can scale without eroding trust.

This is a moment that calls for conviction. Owning the AI Marketing Operating System is not a technical assignment. It is a leadership decision about how growth will be engineered, how capital will be allocated, and how marketing earns its place as a driver of enterprise performance. CMOs who step into this role will not only redefine marketing. They'll shape how their organizations engage with customers to compete in an AI driven world.

Leadership Review

1. Are we leading marketing as a system engineered for growth, or as a collection of functions?

2. Where are we still treating AI as a productivity layer instead of embedding it into workflows?

3. Are teams aligned around a shared vision for reimagining marketing with AI, or optimizing within silos?

4. Do leaders track both performance outcomes and system maturity?

5. Which responsibilities remain owned by the CMO, and which are enabled through Marketing Engineering?

7

Structuring Marketing for Intelligence

*How teams evolve when AI
becomes part of daily operations*

"I have about a hundred people on my global team," a CMO told me. "This covers Product Marketing, Programs, Content, Research, Segment Marketing, and Operations. Everyone is doing a great job. My problem, however, is every team has a different view on how to use AI. Everyone stays in their lane. I worry about an announcement or something from the field on how competitors are using AI to their advantage. That really keeps me up these days."

That comment stuck with me because it reveals a truth many leaders feel but rarely discuss openly. On paper, their organizations look solid. Key disciplines are staffed. Reporting lines are clear. Agencies are identified. The org chart aligns with the business and the work gets delivered. Yet beneath all that, accountability for how the operating model evolves with AI remains undefined. The team has grown by addition with new roles, new capabilities, and new partners, but it hasn't yet accounted for the demands and benefits of artificial intelligence. And as consumer behaviors change, or competitive actions shift, this concern rapidly intensifies.

This is a pattern we see across industries. As noted in Chapter 2, each new era of marketing left behind its own departmental footprint.

Each wave added new channels and roles, but almost nothing was taken away. The result is an org design frozen in time, a multidisciplinary structure that reflects past priorities more than current realities.

The risk lies in assuming that a traditional marketing organization can simply absorb AI as another layer.

The same functions that managed quarterly campaigns and sequential handoffs are now expected to manage machine learning, AI agents, and systems that learn continuously. Traditional marketing team structures reflect a different level of velocity, intelligence, and interdependence.

Think about this through the lens of design. The org chart still reflects span of control and functional boundaries rather than flows of intelligence and work. Accountability sits in vertical silos while modern performance is created horizontally across systems. The very architecture of the team resists the adaptability leaders say they want.

The consequence is persistent friction. Decisions slow down in reviews. AI pilots stall at the edges. Coordination consumes more time than creation. Talent spends more energy navigating the team structure than elevating performance through it. And while teams are working harder, the enterprise feels it in different ways. Costs rise. Go-to-market actions take longer. Platform investments fail to translate into meaningful scale. So what looks like a resourcing problem on the surface, is often an organizational structure problem beneath it.

For marketing leaders, there's almost always a resource gap between what needs to be done and what can be done. That's a given we all live with. And with recent economic volatility, that gap has expanded. That's why CMOs feel the additional weight now. It's not a deficit of skill or ambition. It's a structure problem. And this is where your leadership becomes decisive.

The operating system for marketing requires an organization built for current economic and technological realities. What built marketing's past must evolve to sustain its future. The CMO must own redesigning how skills, workflows, AI agents, and data enable intelligent performance. The solution must adapt to the new realities of economic resources and technology acceleration.

The new operating advantage is the ability to reimagine not only what the team delivers, but how the team is designed to deliver it.

The CMOs who embrace this priority will set the terms of growth for their enterprise. And for CEOs, the implication is that marketing design isn't just an internal debate about roles and workflows. It's a new financial lever that determines margin efficiency, speed of growth, and the ability to scale without adding disproportionate cost. For both leaders the practical question is, are we organized to build and run AIM OS?

When Organizational Design Becomes Advantage

For most of its history, marketing relied on organizational design to manage work, not to drive it. Departments were carved out to contain specialties and the structure reflected control rather than coordination. This made the org chart a map of ownership versus a roadmap for movement. That logic made sense when marketing moved slowly. Annual planning cycles, block media buys, predictable funnels, and annual brand campaigns all rewarded stability over adaptability. The traditional team design protected consistency, but it doesn't optimize coordination, leverage, or speed. This is why the modern operating structure is becoming a performance lever like budget and talent.

AI has changed how marketing operates, so organizational design must change with it.

In modern marketing, the organizational structure is no longer just an Ops responsibility. The structure now functions as the operating platform for AIM OS. It determines what gets done, how fast, by whom, and with what degree of intelligence. It shapes ownership and how insights surface, how resources are allocated, how campaigns activate, and how feedback moves into future action. When the design of the organization isn't built for this level of intelligence and motion, performance plateaus even when talent and budgets are strong.

Marketing today requires an operating model built for flows rather than lines, supporting signal-based execution instead of calendar-based planning, enabling teams to act with foresight rather than convene with hindsight, and most critically, integrating AI as an operating layer of the work.

AI agents are already embedded in workflows such as scheduling campaigns, testing content, tagging assets, predicting next best actions, segmenting audiences, and optimizing spend. But these capabilities create value only when surrounding workflows are designed for speed and decisioning. This includes guardrails that keep the work brand safe and privacy compliant. A traditional organization can't simply "make room" for agents, it must be designed to work with them so that teams and AI agents operate as one system.

Marketing no longer competes on ideas alone.
It competes on the design and health of its operating model.

The most advanced organizations are already moving this way. They organize around intelligence-based capabilities rather than functions. They connect decision makers directly to market signals rather than to reports. They strip out layers that slow the movement from insight to action. And they embed AI into execution not as a pilot but as an operating layer. What emerges is a smarter, better aligned function. A modern operating model is designed for how modern marketing truly performs in real time and in continuous motion.

All of Marketing Must Evolve

As marketing's operating model evolves, every function must refine how it creates value within the system. The temptation in any transformation is to start with the visible edge, like campaigns, channels, content. These are closest to the customer and under the most pressure to deliver. A lot of early generative AI work has taken root here. Yet focusing change only at the edge misses the deeper reality.

Modern marketing performance isn't generated by any single function. It comes from a system of interdependent data, content, decisioning, delivery, and measurement capabilities working together. Said another way, *all* of marketing is affected. Every discipline. Every team. Everyone now contributes to how value is created, delivered, and measured inside the AIM OS. Here are a few examples to illustrate this shift:

- *Brand marketing* must shift from static storytelling to dynamic content frameworks where narratives adapt in real time to market signals.

- *Product marketing* must move from quarterly positioning to continuously tuned messaging (and features) aligned with live buyer behavior.

- *Growth marketing* must evolve beyond nurture calendars toward AI assisted targeting that guides prospects through signal based, personalized journeys.

- *Experience marketing* must become engines of feedback, upsell, and retention, powered by data that flows directly back into decision systems.

- *Creative and content* must shift from asset production to orchestration, designing modular ecosystems that teams and AI can assemble in motion.

- *Market research* must move from retrospective reports to continuous intelligence, providing live inputs that fuel decisioning, personalization, and strategy in motion.

What connects these roles today isn't their place on an org chart or in a funnel. It's the flow of intelligence moving through the system. Every role is now committed to the same flow of intelligence. The breakthrough is created by momentum. Results compound when the system runs as a continuous flow where learning accelerates execution and execution generates more learning. Cross functional complexity is converted into coordinated execution.

Performance gains come from designing conditions where every discipline contributes to momentum. And for the individual marketer wondering what this means for them, the answer is simple. This isn't about replacing roles. It's about reshaping how value is created across the system. Your work and expertise do not disappear, it evolves and scales. Campaign planners become journey engineers. Analysts become decision accelerators. Content creators become orchestrators of modular ecosystems that move in real time.

> *Marketers will not only remain essential,*
> *but they'll also become more impactful.*

The team's expertise will no longer be confined to a channel, campaign, or deliverable. It will be amplified by systems that extend their reach, accelerate their decisions, and scale their creativity. Talent isn't displaced by AI. It is elevated by it. AI agents handle the repetitive and routine, while managers set direction and connect work to strategy. The marketers who lean into this shift will spend less time pushing work through bottlenecks and more time shaping outcomes and guiding the system instead of grinding against it.

Modern marketing is where the role of the individual grows, and where judgment, creativity, and strategic instinct matter more than ever because the AIM OS now carries the load that once slowed them down.

AI is the Operating Fabric of Modern Marketing

AI is the most disruptive technology to enter marketing in decades. It provides the connective layer to supercharge the function. Generative AI aligns with the creative dimension of marketing, accelerating how creative, content, offers, and experiences are produced. While agentic AI aligns with the analytical and reasoning dimension of marketing.

AI is creative scale and the analytical brain, working together to supercharge marketing performance like we've never seen it.

In the future direction of marketing, there's no new architecture without AI. Imagine trying to run marketing today without the internet. It would be unthinkable. The same will soon be said of artificial intelligence.

Most organizations already have concentrated on Generative AI. New platforms and features have appeared, including shadow tools. Useful yes, but they're often applied as a supplement to traditional marketing rather than enabling a new operating model. And with Agentic AI, decisioning is already shifting toward responsibility-based capabilities. Agents take signals, apply rules and models, and execute steps across the system. Marketers set goals and guardrails, and agents do the work and report back. This means that marketing activation is increasingly becoming autonomous, not just automated.

Agentic AI isn't some monolithic robot. It is an ecosystem of specialized AI capabilities, each designed to take on a different slice of marketing work. For the modern CMO, the scope of leadership now extends beyond people and partners to include these AI agents. Managing and orchestrating both people and AI agents is an important part of modern marketing leadership.

Exhibit 12: Common Marketing Agent Responsibilities

Each agent has a single primary responsibility. Real value emerges from how they operate together, and how they are orchestrated with system intent.

Area	Agent	Role in the Marketing System
Customer & Market Understanding	Customer signal agent	Detects shifts in customer intent by interpreting live behavioral and transactional signals across channels.
	Audience formation agent	Builds and updates audience segments dynamically so targeting and personalization stay current.
	Market sensing agent	Monitors competitive and category signals to inform strategic marketing decisions.
	Customer feedback agent	Aggregates sentiment, reviews, and qualitative feedback to surface emerging customer themes.
Strategy & Planning	Growth alignment agent	Aligns marketing actions to defined business objectives so execution stays tied to growth priorities.
	Next best action agent	Selects the optimal message, offer, or experience based on real time decision logic.
	Experiment design agent	Models scenarios and prioritizes tests to evaluate impact before actions are activated.
	Resource allocation agent	Adjusts budget and effort dynamically to improve efficiency and outcomes.
Brand & Creative	Brand integrity agent	Ensures all generated and automated outputs remain consistent with brand standards.
	Creative variation agent	Produces and adapts creative variants within approved brand and compliance guardrails.

	Content planning agent	Determines what content is needed based on customer intent and journey stage.
	Message coherence agent	Maintains consistent value propositions and messaging across channels and journeys.
Campaign & Journey Execution	Campaign assembly agent	Assembles and activates campaigns in real time using modular assets and orchestration logic.
	Journey orchestration agent	Coordinates multi-step customer journeys that adapt to behavior as it unfolds.
	Channel execution agent	Executes marketing actions within specific channels according to system instructions.
	Execution telemetry agent	Captures real time delivery and engagement signals to provide visibility into execution health.
Performance & Optimization	Performance monitoring agent	Analyzes engagement and outcome patterns to identify opportunities for improvement.
	Conversion optimization agent	Adjusts experience logic and sequencing to improve response and conversion rates.
	Attribution intelligence agent	Determines which actions contribute to outcomes across complex customer paths.
	Reuse identification agent	Identifies high-performing assets and logic that can be reused and scaled.
Governance, Trust & Control	Decision audit agent	Reviews data flows and decision logic to detect anomalies or unapproved behavior.
	Privacy & compliance agent	Ensures automated actions adhere to consent, privacy, and regulatory requirements.
	Brand safety agent	Evaluates outputs for brand, legal, and ethical risk before and during activation.
	Bias detection agent	Identifies unfair or harmful outcomes and escalates them for management review.

How Structure Becomes the Difference

If AIM OS defines how intelligence runs across the function, then structure defines who makes it work. Most CMOs already have the core marketing talent on the team. Analysts, researchers, campaign managers, brand leaders, and specialists in content or data are in place. What is missing is how these people are connected. Without a new design which has clear ownership, talent stays trapped in silos, insights stall in isolation, and technology underdelivers against its promise. Marketing teams now compete on the design of their capabilities and operating models.

Exhibit 13: **Legacy Operating Silos vs. AIM OS Performance Zones**

Dimension	Legacy Silos	Performance Zones
Accountability	Fragmented by function, deliverables planned	Ownership by outcome, embedded in design
Decision Speed	Sequential reviews, scheduled cycles	Continuous flow, real time prioritization
Data & Insight	Isolated in data silos and reports	Circulates across loops, feeds directly into action
Execution	Manual handoffs, coordinated bottlenecks	Orchestrated workflows, AI assisted output
Cost Profile	More headcount needed to scale	System leverage multiplies capacity without cost
Trust & Governance	Added at the end of workflows, reactive checkpoints	Built in, automated guardrails, proactive oversight
Resilience	Fragile, slow adaptation to market shifts	Adaptive, system learns and evolves continuously

This table shows how performance is unlocked when marketing is organized around the four capability zones: Adaptive Decisioning, Autonomous Activation, Experience Loop Intelligence, and Intelligence Governance. These form the system level structure of modern marketing.

Each capability zone integrates traditional roles that already exist in most organizations, while integrating Marketing Engineers and Agent roles that make the system run. And when these zones align, marketing stops functioning as a collection of departments and starts operating as a cohesive, modern intelligent organization.

Zone 1: The Adaptive Decisioning Team

Adaptive Decisioning is the capability where data signals become strategy. This zone shifts decision making from rooms to flows. Instead of waiting for the next planning cycle, marketing decisions are guided by embedded logic, predictive models, and rules that turn a broad range of data into action.

The system defines in advance which signals trigger which actions and how resources are allocated. Decision architects own the logic. Model builders ensure AI aligns with marketing context. Business leaders still set strategic direction, but operational choices happen inside the workflow under specific guardrails.

This shift reduces bottlenecks and makes accountability explicit. Organizations that design this zone deliberately reduce time to decision, cut campaign lag, and ensure that intelligence drives desired outcomes.

Traditional Role Alignment: Current marketing analyst roles include Marketing Data Analyst, Customer Insights Analyst, Market Research Manager, Product Marketing Manager, Marketing Operations, Competitive Intelligence Analyst, and Campaign Analyst.

Marketing Engineering Roles: Examples of some of the new roles associated with engineered reasoning and decisioning in this zone:

- AI Decision Architect defines decision frameworks and embeds prioritization logic into workflows.

- Model Translator converts model output into business-ready rules.

- Signal Orchestrator routes high-value signals into systems where they can trigger action.

- Adaptive Targeting Specialist designs AI driven targeting logic to ensure precision at scale.

- Predictive Intelligence Lead owns forecasting models and integrates them into go-to-market execution.

- AI Marketing Strategist connects business goals with AI powered decision flows.

Aligned Marketing Agent Roles: AI agents now work alongside these roles, automating reasoning and routing tasks once handled manually. They power the decisioning by interpreting signals, evaluating options, and selecting actions before execution begins. They compress the distance between insight and choice while operating within defined constraints.

- Customer signal agent interprets behavioral and transactional signals to detect shifts in customer intent.

- Audience formation agent dynamically builds and updates segments and intent states based on real time signals.

- Next best action agent determines optimal messages, offers, or experiences based on decision logic and thresholds.

- Resource allocation agent optimizes budget and effort allocation in response to performance and demand signals.

- Experiment design agent models scenarios and prioritizes tests to evaluate potential outcomes before activation.

Together, these agents act as extensions to the decisioning team. They maintain constant situational awareness, reduce latency between signal and response, and free talent to focus on strategy and creativity rather than mechanics.

Why It Matters: When Adaptive Decisioning is designed well, it closes the gap between signal and response. Program leaders see campaigns adapt in real time. Customer data teams see clean events streaming into models. And analytic teams see decisions instrumented. Marketing shifts from after-the-fact dashboards to predictive execution in the flow. The impact is measurable: signal-to-action time falls, decision-automation rate rises, and faster decision cycles lower customer-acquisition cost as spend moves away from low-propensity paths. Targeting and prioritization lift lifetime value by routing customers into higher-fit journeys. Throughput increases as less time goes to coordination and more to shaping outcomes.

Adaptive Decisioning in Practice: At a global SaaS company, campaign decisions were slow and disconnected from customer behavior. Analytics produced reports, but those insights rarely reached frontline activation. Campaigns ran on monthly sprint calendars while AI models remained unused. The CMO created Project IQ, an Adaptive Decisioning Team of marketing engineers designed to create autonomous choices. The team included a Decision Architect translating strategy into prioritization rules, a Data Analyst surfacing live intent signals from usage and engagement, a Journey Strategist mapping flows into decision logic, and a BI Engineer tracking model performance.

Together, the team defined triggers for demo offers and routed next-best-message logic directly into the activation platform. Within two quarters, time to decision dropped 40 percent. Customer-acquisition efficiency improved as prospecting became more precise. AI shifted from shadow pilots to system inputs, and marketers began treating IQ as a central part of the go-to-market workflow.

Zone 2: The Autonomous Activation Team

Autonomous Activation is the capability where marketing executes at scale through engineered orchestration. Traditionally, campaign managers, channel leads, and content teams drove execution through handoffs and production queues. And yes, CMOs frequently comment that even with automation platforms, their workflows remain fragmented.

In modern marketing, execution is orchestrated. Manager strategies shape the logic, and agents do the heavy execution lifting. Traditional roles now act as system operators, supported by marketing engineers who elevate AI driven workflows and performance.

Traditional Role Alignment: In most organizations, activation relies on Campaign Operations Manager, Digital Marketing Manager, Social Media Manager, Content Manager, Thought Leadership Coordinator, Marketing Automation Specialist, Performance Marketing Manager, Email Marketing Manager, and Channel Marketing Lead.

Marketing Engineering Roles: New titles reflect the shift to systems and orchestration:

- Activation Engineer designs and manages AI first execution systems.

- Agentic Orchestration Manager oversees autonomous agents that drive campaigns.

- Journey Automation Lead builds always-on, adaptive workflows.

- AI Content QA Lead ensures quality, brand compliance, and AI readiness in content at scale.

- AI Strategy and Deployment Manager align AI solutions with execution pipelines and scale needs.

- AI Marketing Strategist bridges strategy and execution, weaving AI activation into go-to-market programs.

Aligned Marketing Agent Roles: AI agents execute the operational rhythm of marketing activation. They transform strategy into coordinated actions that run continuously across channels and contexts. All execution is governed by predefined guardrails and real time visibility.

- Campaign assembly agent reads campaign logic, selects modular assets, and activates messages across platforms in real time.

- Creative variation agent generates and adapts copy, imagery, and variants within approved brand and compliance constraints.

- Channel execution agent delivers actions within specific channel platforms according to orchestration logic.

- Execution telemetry agent captures real time delivery, engagement, and response signals to provide visibility into execution health.

- Brand safety agent evaluates outputs against brand, legal, and ethical standards before and during activation.

Together, these agents enable marketing to operate at speed and scale without sacrificing control, ensuring execution remains aligned to intent while producing the telemetry data required for learning and governance.

Why It Matters: Autonomous Activation turns program execution from manual pushes to system-driven orchestration. Launch briefs become logic, assets become modular building blocks, and channels are coordinated. Program leads evolve into orchestration leaders who design the rules and tune the system, while engineered activation roles keep it reliable and scalable with built-in guardrails that keep the work brand safe, privacy compliant, and outcome aligned. ROI increases as your team executes more targeted experiences, tests faster, and scales without extra handoffs.

Autonomous Activation in Practice: At a national retailer we worked with, campaign launches stalled under manual briefs, approval bottlenecks, and siloed logic. Automation investment failed to meaningfully increase reach or scale. The CMO expanded a generative advertising pilot into an Autonomous Activation Team of marketing engineers to orchestrate execution. The structure included a Journey Automation Lead, a Content Ops Manager empowered by generative AI to produce modular assets, a Channel Engineer managing trigger logic across channel programs, and an AI Performance Lead monitoring autonomous agent behavior and tuning model performance.

The system chose timing and offers in real time. Context engineers tuned prompts, enforced brand rules, adjusted guardrails, and built reusable campaign programs. Campaign cycle times dropped 65 percent. Content output scaled fivefold, and segment testing volume increased threefold.

Zone 3: The Experience Loop Intelligence Team

Experience Loop Intelligence is the capability that converts feedback into continuous learning. For decades, teams focused on execution and measured the performance of programs after the fact. These efforts were sequential, disconnected, and slow. In a modern structure, insights, data, research, and analytics are fused into a continuous feedback system. Every touchpoint now becomes both engagement and learning input. Intelligence from journeys, and personalization engines flows directly back into the Adaptive Decisioning Zone. Now, learning velocity becomes a performance driver, and the faster the feedback, the smarter the system.

Traditional Role Alignment: Current roles in this zone include Customer Experience (CX) Manager, Journey Analyst or Strategist, Persona Segment Manager, Market Research Analyst, Customer Insights

Analyst, and Digital Experience Manager or Personalization Specialist.

Marketing Engineering Roles: Modern teams are adding roles to manage real time learning:

- Program Engineer integrates martech, AI, and experience execution across channels.

- Optimization Engineer builds and operationalizes adaptive testing and feedback systems.

- Signal Integration Lead unifies diverse customer signals across platforms into actionable inputs.

- AI Experience Orchestrator designs dynamic, adaptive customer interactions.

- Marketing Data Engineer owns the speed and quality of learning cycles.

Aligned Marketing Agent Roles: AI agents extend the experience intelligence by closing the distance between sensing and learning. They interpret feedback, detect patterns, and adjust experience logic so performance improves through use.

- Performance monitoring agent analyzes engagement and outcome patterns to identify deviations and opportunities for improvement.

- Feedback agent aggregates sentiment, reviews, and qualitative feedback across channels to surface emerging themes.

- Conversion optimization agent adjusts experience logic, content selection, or sequencing based on predictive response and observed behavior.

- Customer signal agent converts unstructured inputs such as social posts or transcripts into structured signals usable by analytics and decisioning.

- Learning retention agent validates outcomes and routes confirmed insights back into models and workflows so future actions reflect what the system has learned.

Together, these agents ensure learning compounds rather than resets. They turn fragmented analytics into active learning loops to increase the system's ability to adapt in near real time. This allows experiences to adapt continuously as customer behavior evolves.

Why It Matters: Experience Loop Intelligence engineering builds resilience by enabling campaigns to evolve in real time, messaging to adapt to sentiment, and journeys to refine continuously. When traditional roles are embedded into these feedback flows, they gain greater influence over buying decisions rather than simply delivering outputs. Engineered roles keep the system learning and adapting, with guardrails that protect brand integrity and privacy without slowing the work.

Experience Loop Intelligence in Practice: At a global executive recruitment firm we consulted with, customer feedback data had accumulated in surveys but produced little insight. Journeys were built on assumptions, and testing models were almost absent. We helped them close the gap between campaigns and recruiter contacts.

The marketing engineers established feedback flows, so signals drove change in real time. When new thought leadership (TL) asset downloads dipped below benchmarks, message shift rules activated. Content adjusted toward different topics and drivers, and the journey orchestration reconfigured automatically. Program leaders could see which triggers fired and why. Customer data teams streamed clean events and features into models, and analytics traced decisions through explainable logs. This was engineered intelligent system design in action. Test velocity tripled, recruiter outreach rose 22 percent, and average response time to feedback fell from four days to two.

Zone 4: The Intelligence Governance Team

Intelligence Governance is the capability that makes intelligence on brand, trustworthy, and explainable. In traditional marketing operations, governance came at the end. Brand teams fixed tone, Legal reviewed content, and IT checked compliance. All of this is necessary but often reactive.

In a modern marketing engineering structure, governance is embedded in the workflows themselves. Brand, Legal, Compliance, and IT stakeholders are integrated directly into the AIM OS. Governance becomes a design principle instead of a checkpoint.

Traditional Role Alignment: Roles already involved include Brand Manager, Brand Compliance Lead (often referred to as "Brand Police"), Creative Director, Legal and Compliance Manager, Data Privacy Officer, Martech Platform Owner, Quality Assurance (QA) Manager, and Corporate Communications Manager.

Marketing Engineering Roles: New roles reflect how trust and accountability now operate as system-level capabilities:

- Governance Engineer embeds controls, auditability, release gates, and risk checks into AI workflows.

- AI Model Auditor monitors AI outputs for bias, drift, and compliance.

- Ethical Targeting Specialist ensures personalization follows ethical and privacy standards.

- Trust and Transparency Manager builds visibility into system decisions and explainability trails.

- Adaptive Governance Lead manages evolving governance logic in real time, enabling self-governing systems.

Aligned Marketing Agent Roles: AI agents embed governance directly into the marketing system, validating that decisions, content, and workflows operate within defined boundaries. They flag issues before they create exposure. Oversight becomes continuous rather than episodic.

- Decision audit agent reviews data flows, decision logic, and model outputs to detect anomalies, drift, or unapproved behavior.

- Privacy and compliance agent validates that automated actions adhere to consent, privacy, and regional regulatory requirements.

- Brand integrity agent monitors tone, message alignment, and visual execution to ensure brand standards are upheld across outputs.

- Bias detection agent identifies unfair or harmful outcomes and escalates them for team review.

- Transparency agent produces explainability summaries and dashboards that provide visibility for executives and regulators.

These agents act as embedded safety systems that balance automation with accountability. They ensure that intelligence runs fast but stays inside trusted boundaries, turning governance from a bottleneck into a performance enabler.

Why It Matters: Governance is the structural foundation for decisioning and automation at scale. For CEOs, it makes performance visible and auditable. For Legal and Privacy, it builds consent, data use, and recordkeeping into the workflow. For Brand and Social, it encodes voice, tone, and escalation rules so content stays on brand even at speed. Marketing engineers wire these controls into AI systems, so speed and safety coexist with clear visibility. This enables marketing teams to advance the AIM OS while managing trust and compliance exposure.

Intelligence Governance in Practice: Working with a consumer goods company, an AI program platform was adopted quickly but without guardrails. Autonomous agents pushed content inconsistently, brand voice and tone drifted, and manual oversight and sampling increased. Leaders worried trust would erode.

The corporate marketing team responded by codifying the human-in-the-loop process, creating an Intelligence Governance Team that encoded brand standards, privacy rules, and release gates into the system. This gave the team auditable visibility, Legal and Privacy built-in compliance, and Brand and Social teams on-brand execution with defined escalations.

The structure included a Model Ops Lead inserting compliance guardrails into generative workflows, a Brand Governance Lead coding guidelines directly into AI prompt libraries, and a Platform QA Manager embedding automated safeguards into CMS, CRM, and marketing tools.

This was engineered governance the team now affectionately calls "The Bureau." Automated QA flagged tone deviations, shared brand prompt libraries, and audit logs enabled traceability. Governance is about engineering trust into the system. Marketing engineers build the controls that make AI systems fast and auditable.

Summary

Leading marketing in the AI era requires more than adjusting org charts or creating centers of excellence. The scale of change now underway calls for a re-architecture of how marketing works.

Modern performance is created through collaboration. Traditional marketing brings judgment, creativity, and deep market intuition. AI agents extend that expertise with speed, scale, and continuous execution. Marketing Engineers design the operating system that integrates these strengths into a reliable whole. Together, these capabilities reinforce one another, allowing performance to compound as insight, execution, and learning move in sync.

The future of marketing is defined by zones of performance aligned to the AIM OS. Each zone represents a capability required to compete in an AI driven marketplace. These zones cut across traditional job descriptions because modern performance cuts across traditional workflows. In practice, they become the organizing logic for how people, platforms, data, and AI work together to drive results.

This shift changes the questions marketing leaders ask:

- From: Who owns this deliverable?
  To: Which zone is accountable for this outcome?

- From: How do we pass this insight to another team?
  To: How does the system feed this back in real time?

- From: Do we have enough headcount to execute?
  To: How do we scale activity through AI and automation?

The effects are significant. Efficiency improves and resilience rises as the system adapts to market change. Accountability is designed upfront, creating clarity instead of negotiation. Trust scales because governance is structural and embedded. Growth accelerates as automation increases output without multiplying headcount.

Operating design becomes the roadmap for how marketing gets done. It determines how signals flow into decisions, how decisions trigger action, how resources are allocated, and how outcomes feed learning back into the system.

Organizing around capability zones shifts work through flow rather than hierarchy. Data circulates instead of sitting in silos. Decisions become continuous rather than scheduled. Roles expand in impact. Marketers focus on strategy and creativity. AI agents handle scale and execution. Marketing Engineers ensure the system performs as intended.

This is the difference between running functions and operating a system. The CMO makes that evolution explicit by multiplying impact through design. The strongest marketing organizations thrive because expertise is extended through intelligence and reach, while culture and operating design advance together.

For CEOs, the implication is team design becomes a financial lever. A marketing organization built for flow drives faster acquisition, higher conversion efficiency, and lower execution cost. Instead of incremental lift from isolated campaigns, the enterprise gains compounding performance.

Leadership Review

1. Do we have clear ownership for each performance zone with a quarterly roadmap?

2. Are workflows engineered for speed, decisioning, and throughput, or slowed by reviews and approvals?

3. Have we appointed a Chief Marketing Engineer to own AIM OS and its evolution?

4. Are Marketing Engineers embedded across zones, or are we still hiring traditional roles?

5. Is AIM OS designed to route signal to action intelligently, or is it fragmented and manual?

8

Escaping the Grip of Legacy Marketing

*Why progress often stalls inside
even the best teams*

A CEO I met with recently was quick to comment on their AI adoption progress. "We've been at it for twelve months and I haven't seen any real impact on growth or productivity yet." Most executives I hear from don't say this, so I asked, "What has changed about the way marketing runs?" He paused. "I hadn't thought of it that way," he admitted, and then asked "Wait, so how is AI changing marketing?" That was the issue. I learned later that the company had been busy experimenting, but the operating model hadn't evolved. Pilots were stacked onto old workflows, activity multiplied, and performance stayed the same.

Many organizations fall into the same pattern. They treat AI adoption as a move from old to new, part legacy and part modern, where progress feels tangible. Teams adopt AI tools, launch pilots, log quick wins, and watch activity rise, yet the structure stays the same. Data continues to sit in silos. And programs continue to follow the calendar rather than the customer.

AI's promise is real, but it places CMOs in a difficult position. They must run the traditional marketing machine while simultaneously building an AI driven operating model. Both demand leadership. Both consume resources. And they rarely move at the same speed.

This tension is where the inertia takes hold and creates the illusion of progress. Leaders try to keep both sides running, driving the legacy machine at full speed while stacking modern capabilities on top. The result is a more expensive version of the old model. From the outside it looks promising. From the inside it feels exhausting. It becomes a system stretched between old logic and future demands. In truth it delivers the worst of both worlds with new intelligence capabilities layered on slow, manual, department bound processes.

Breaking free comes from making harder choices.

It means reallocating resources, redesigning workflows, and elevating modern marketing to a top line business priority. Treat it as a side project and it stalls, because the power of 'business as usual' always wins out.

Competitors who rewire their systems redesign their operating models rather than layering AI onto legacy structures. They move faster, learn faster, and compound performance long before others recognize what has changed. The reality is the middle ground offers no stability. Partial transformation isn't a stage. The longer leaders remain there, the harder it is for your team to escape.

As shared in the Case Study, leadership conviction often matters more than internal discomfort. Marketing's role isn't to trail the business. It's to help define where the organization must go next.

Case Study: Strategic Conviction and Market Timing

At Cognizant, the leadership team made an explicit five-year strategic commitment to grow the company's digital transformation consulting and technology services. This shift required expanding capabilities. It also required repositioning the brand from an IT service provider to a business advisor that understood how companies operated and how digital technologies could reimagine those operations. Marketing was tasked with leading that repositioning.

Our corporate marketing team worked closely with business units to establish a new portfolio naming architecture, introduce new brand messaging, scale thought leadership, expand digital channels, launch targeted television, and host marquee thought leadership client events. The goal was to reach new decision makers and establish Cognizant as a leader in a category that was still taking shape.

As the repositioning gained traction, some business leaders raised concerns that brand perception was advancing faster than new capabilities were being delivered. The question surfaced whether marketing and communications should slow down to better match the pace of execution inside the business.

At that moment, a critical leadership decision was made. Malcolm Frank, the Chief Strategy Officer and Chief Marketing Officer, discussed the situation directly with Frank D'Souza, Cognizant's CEO. D'Souza understood both the importance of the market opportunity and the reality that clients were looking to Cognizant for guidance on digital transformation. His direction was that marketing should continue accelerating the brand's association with digital transformation, and the business would move faster to meet the market expectations being created.

Frank's commitment to the strategic direction was pivotal. This set the organization in motion, accelerating digital revenue growth, strengthening customer confidence, and reinforcing investor confidence in the strategy and execution. The decision reflected a core leadership principle. When strategic vision is clear, leadership conviction matters more than internal discomfort.

Cognizant's experience illustrates why making the shift to AIM OS requires CMO conviction. Customers don't wait. Markets won't pause. Leaders who hesitate risk surrendering relevance and competitive advantage to competitors willing to lead, and whose systems continuously learn.

The Four States of AI Marketing Transformation

For every CMO, this is a defining inflection point. Organizations caught in the middle are standing at the edge of a new performance opportunity. The leaders who step forward will modernize marketing.

Most marketing organizations were built for a different set of demands. When leaders pursue modernization, they often add new tools onto the traditional operating model. Activity increases while the system remains unchanged. Transformation gravitates back to familiar areas, and progress stalls. Cognizant avoided the stall by refusing to let internal pacing dictate market positioning. A single decision kept the organization out of the middle.

We see this pattern in nearly every engagement. Leaders often know their ambition and direction, but they don't always know the actions or behaviors their team needs to change. They point to new technologies as proof of progress while the underlying operating model remains intact. That's why the AIM OS transformation journey map has become essential. It helps teams see when they are drifting and clarifies the path toward engineered marketing performance.

Exhibit 14: **AIM OS Transformation Journey Map**

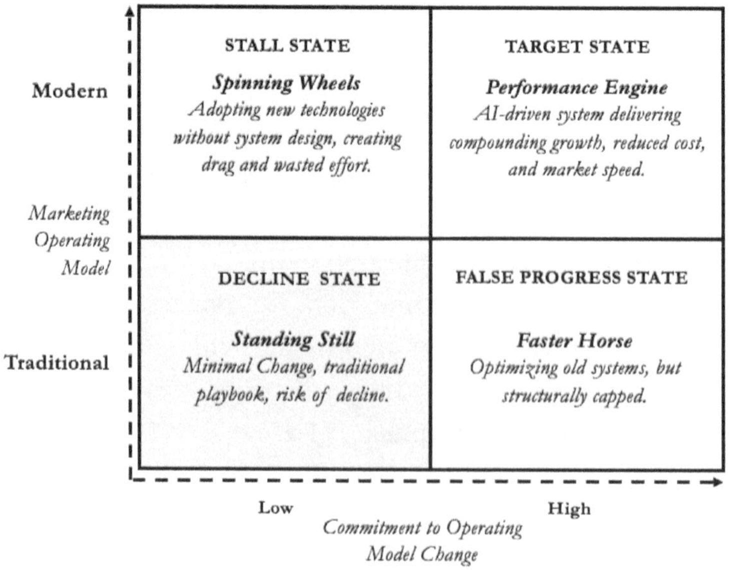

Every AI marketing transformation sits within one of four states. Some leaders cling to the legacy playbook. Others get trapped in urgent activity or false progress. But leaders who break through, redesign their operations for real speed and scale. The value of this map isn't labeling where you are, but in recognizing the drift and knowing what it takes to reach the AI Marketing Operating System state.

Bottom left: *Standing Still*, the Decline State

The legacy playbook. Campaigns planned by calendar, approvals moving through email and meetings, data locked in silos. Teams may feel steady, but they are already losing ground in the AI era. *Lesson:* if your workflows look the same as three years ago, you aren't standing still, you are sliding backward.

Bottom right: *Faster Horse*, the False Progress State

The trap of apparent momentum. Cost pressures accelerate, platforms and dashboards multiply, workstreams increase, but the operating rhythm hasn't shifted. Leaders here optimize the old machine instead of redesigning it. *Lesson:* optimizing and scaling legacy methods isn't transformation. It only reinforces the old model. Before making a major investment, ask whether you are building for the future or fine-tuning the past.

Top left: *Spinning Wheels*, the Stall State

The most common trap. Teams leave the legacy model, add modern tools and roles, and become busier than ever, yet the operating model stays the same. Calendar-based campaigns, matrixed execution, and fragmented data persist. It is neither traditional nor modern, too costly to be the old model, too slow to be the new one. *Lesson:* if your marketing feels both modern and manual with the same intensity, you aren't advancing, you are stuck in the middle.

Top right: *Performance Engine*, the Target State

The destination state. Workflows are redesigned for speed. Data flows seamlessly. AI is embedded in execution. The operating model doesn't just look different, it runs differently. *Lesson:* when marketing feels more orchestrated, like an engine in motion, you have crossed into engineered performance and are running AIM OS.

For CMOs and CEOs, the value of this map is in reading the behaviors that appear in everyday marketing work. The Decline State shows when legacy habits are pulling you backward. The False Progress State reveals when investments anchor you to a legacy model. The Stall State exposes the illusion of being both modern and manual at once. Only the Target State compounds value and builds competitive advantage. Used this way, the map becomes more than diagnostic. It's a compass that helps leaders recognize drift early, avoid false momentum, and steer decisively toward engineered performance.

The Compounding Cost

When marketing remains caught between old and new models, expenses rise in subtle ways, through incremental headcount, layered technology, and growing reporting effort. What appears manageable in the short term gradually undermines efficiency and weakens the economic foundation of performance.

Survey data confirms the pattern. In our leadership study, 77 percent of CMOs report operations marked by complexity and inefficiency. Only 11 percent believe they are fully leveraging their platforms. Just 16 percent report data infrastructure capable of supporting AI effectively. Together, these findings point to structural constraints rather than execution gaps.

Hidden costs compound the impact. Teams lose time reconciling mismatched data. Campaigns slow as approvals queue up. Reporting cycles stretch until decisions are based on stale data. As complexity grows without the right operating model, more energy goes into keeping an outdated system running than improving performance. Talent that should be driving growth is redirected toward maintenance instead of momentum.

Most concerning, high performers burn out chasing workarounds and waiting on decisions that should move automatically. Over time, momentum slows, morale softens, and attrition increases, with each departure reducing capability. What appears as progress on the surface often masks a gradual loss of capacity and competitiveness. The cost shows up in missed opportunities, slower growth, and the increasing effort required to regain competitive advantage.

Reading the Illusion States: Key Indicators

The hardest traps to recognize are the ones that feel like momentum. Projects launch. Tools go live. Dashboards expand. On the surface there is motion. Look closer and a different pattern emerges. Activity increases while velocity stays flat. If several of these patterns sound familiar, you aren't moving toward a performance system, you are scaling the past.

To understand why this happens so often, it helps to look at the two states that most often surround progress on the path to modern marketing.

Faster Horse (False Progress State): Optimizing Traditional Operations

- *More tools than value*: Stacks grow every year and the slide of logos looks impressive, yet features sit unused, adoption remains shallow, and integrations stop at the basics. New tools launch with fanfare then get shelved because teams "aren't ready" or integration is deferred. What looks like progress is accumulation without capability.

- *Better data, not faster data:* Dashboards become more polished and reports multiply, but they still take days or weeks to produce. Customer signals arrive, then sit idle until the next cycle. Insights are shared in meetings but rarely change what happens in the market that week. Leaders celebrate the volume of data and miss the timeliness of signals, where the real advantage lives.

- *Work still runs on meetings and spreadsheets:* Decks remain the workflow. Calendars fill with status calls. Approvals travel in long email chains. New platforms sit beside old habits instead of replacing them. The organization is busier, but the rhythm of work hasn't changed.

Case Study: The Cost of Optimizing the Past

A global manufacturer set out to expand its account-based marketing program. Regional teams across several countries had built campaigns by hand for dozens of segments. On paper the program looked advanced with tailored messages, market-specific assets, and dedicated production pods.

When leadership chose to scale, they added low-cost marketing-as-a-service capacity instead of modernizing orchestration with AI and automation. Onboarding slowed. Execution stayed manual. Once campaigns launched, there was no way to adjust in real time. They remained tied to a quarterly rhythm.

A competitor chose a different path. AI personalization ran at the individual level in minutes instead of weeks. The system tested, learned, and refined daily. Share of market increased. *Lesson:* optimizing and scaling a legacy method isn't transformation. It's entrenching the old model, and it creates a weaker position.

Spinning Wheels (Stall State):
Declaring AI Without Structural Change

- *Automation in pockets:* Isolated campaigns run with triggers or workflows, but there is no end-to-end design linking decisioning, content, and delivery in real time. Demos look modern, yet the customer experience still feels slow and generic.

- *Talent misallocated:* Analysts reconcile mismatched data. Strategists wait for missing inputs. Developers fix the same errors month after month. People who could shape the future are maintaining the past. Activity gets confused with value and strategic capacity erodes.

- *Performance plateaus:* A new platform creates a temporary bump that fades by the next quarter. Output flatlines. Teams work harder just to hold the line, and the pattern repeats with higher spend and no durable lift.

- *The calendar drives work, not the customer:* Activity follows fiscal cycles instead of live demand. Teams stay busy, but the rhythm is set by an internal process rather than customer signal.

Case Study: When Modernization Becomes a Side Project

A global B2B fintech brand set out to modernize marketing with AI driven decisioning, stronger data integration, and automated workflows. On paper the initiative looked great. In practice it lived inside a small innovation team with limited authority, no direct CMO ownership, and no link to performance metrics. It competed with a dozen other projects drawing from the same budget and the same talent pool.

After eighteen months, the output was pilot reports and internal presentations. Cycle times didn't improve. ROI stayed flat. The daily operating model of marketing looked unchanged.

The CMO of a well-known category competitor led the program directly, built it into the marketing plan, funded it with a dedicated budget, assigned cross-functional leaders, and reviewed progress weekly. By the time the first company's pilot team prepared its third executive update, this competitor had embedded agents into workflows and used freed capacity to increase market testing for new services. Lesson: modernization cannot live as a discretionary project. The CMO must own it with authority, accountability, and resources. Otherwise, the pull of business as usual will always win.

Breaking The Stall Cycle

The stall often begins with tactical success. New tools are added. Workstreams are launched. Roles expand. Each move feels productive in isolation, but the underlying structure remains intact. Over time, all of this motion is mistaken for modernization. But real progress begins only when leaders confront how work actually moves, how data flows, and how decisions are made.

Breaking the stall in the middle doesn't mean asking teams to move faster. It requires changing the structure they are moving within. Organizations that escape it make sharper choices. They stop investing in activity that pulls them backward and redirect resources toward capabilities that compound performance over time.

Exhibit 15: **Stop / Start CMO Decisions**

Decision Area	Stop What Sustains the Stall	Start What Unlocks Modern Performance
Martech Investment	Funding new tools without integrating or governing the stack.	Maximizing return on existing platforms through orchestration and system design.
Resource Management	Expanding headcount to compensate for manual processes.	Automating execution to increase impact without increasing fixed costs.
Operating Model	Running legacy workflows through emails, manual approvals, and handoffs.	Redesigning workflows for data driven reasoning and decisioning.
Data Architecture	Allowing fragmented data to limit speed and foresight.	Building unified AI ready data foundations.
AI Investment	Treating AI as isolated pilots and experiments.	Embedding AI in core workflows where it drives measurable advantage.
Talent Strategy	Hiring to sustain yesterday's ways of working.	Building hybrid, AI fluent teams that scale impact through AIM OS.

Escaping false progress and stall states isn't about working harder or buying more tools. It begins with structural redesign. Marketing leaders who move beyond half-transformed models and commit to system-level change create the conditions for compounding performance. When marketing operates as an intelligent operating system, acceleration becomes natural and momentum is built into how the function learns, adapts, and grows. Remember, breaking free from the past comes from making harder choices.

The Competitive Consequence

Half-transformed systems create visible competitive weakness. Competitors recognize it in your slow response times, uneven experiences, and delays in deploying new capabilities. Customers feel it when your brand is a step behind the most relevant, responsive players.

> *In modern marketing, system learning and*
> *cycle speed are competitive moats.*

Once a competitor operates at a higher cycle speed, every day you remain stalled widens the gap, and every cycle you miss shows up as higher cost of growth.

While you debate priorities or pause for the next phase, competitors with rewired systems are already running a performance system on AIM OS. They don't wait for quarterly cycles. They execute continuously, adjusting offers and creative in real time. They don't debate over which data source is correct. Their systems reconcile it automatically and act within minutes.

Each completed cycle sharpens their advantage. Each test compounds learning. While budgets are being approved for next quarter, these organizations have already optimized multiple journeys, redirected spend toward top-performing audiences, and deployed personalized content at scale.

The reputational impact compounds the damage. Customers don't see your infrastructure, but they do experience its performance. When competitors deliver faster, more relevant, and more consistent interactions, your brand appears slower, even with greater investment. As marketers we know these perceptions form quickly, and it's difficult to reverse.

In short, stalling has real consequences. It creates pursuit rather than stability. In a pursuit, the leader sets the pace, the benchmarks, and the customer's expectations. At that point, competition shifts from winning to catching up.

Case Studies of Leadership Blind Spots Across Industries

Across industries, the same leadership blind spot repeats. Organizations try to operate in two worlds at once. They maintain the old model while layering pieces of the new on top. What looks like progress at first slowly increases cost and slows momentum. Complexity rises. Speed falls. The organization carries the weight of both models without fully benefiting from either. This challenge spans every sector because it's structural.

The confusion often starts when activity is mistaken for acceleration. In retail, seasonal collections were once planned months in advance. Brands like Zara changed the competitive equation by rebuilding their supply chains for speed, shrinking design-to-shelf cycles from months to weeks. Velocity became the advantage. The lesson applies directly to marketing. Calendars and campaigns create motion, but only signal-based systems create sustained speed.

The same pattern appears when tools are mistaken for transformation. Automakers once bolted hybrid components onto combustion platforms, gaining incremental gains but preserving structural limits. The real breakthrough came when companies like Tesla and BYD rebuilt the architecture entirely around electric systems. Marketing faces the same choice. Adding AI to legacy workflows increases complexity. Re-architecting the system is what unlocks performance.

Spend is often mistaken for scale. Traditional banks digitized by adding mobile apps on top of branch-era cores. Digital-native banks rebuilt around cloud-first architectures, enabling faster onboarding, real time services, and lower operating costs. Marketing follows the same logic. Larger budgets don't guarantee reach or relevance. Systems designed for throughput and learning do.

The pattern holds across categories. Operating in two worlds increases cost, fragments effort, and slows progress. Breakthroughs occur when leaders stop layering the new onto the old and redesign the structure so AIM OS can operate at full speed.

Breaking Out

Breaking out of half-transformed systems requires more than effort or new tools. It requires reengineering the marketing system. The goal is a fundamental shift in how marketing senses, decides, acts, and learns continuously, rather than moving through periodic review cycles.

The AIM OS Organizational Health Scorecard clarifies where progress has stalled. It shows whether marketing is operating in Decline, False Progress, or Stall states, or beginning to function as a true performance system.

*Breaking out of the middle requires structural redesign
rather than incremental optimization and change.*

The work happens while AIM OS is running. Progress starts with targeted redesigns in high-impact areas where faster cycles, greater scale, and higher adaptability produce visible business payoffs. The scorecard highlights where to begin by revealing the constraints that hold organizations in False Progress or Stall.

As organizations move beyond these middle states, the signals change. Cycle times compress as campaign-to-market speed drops from weeks to days. Platforms deliver value through deeper orchestration, automation, and personalization rather than expansion. Data stays fresh, with insights moving from collection to action within hours. AI advances from experimentation to execution, embedded directly in workflows where agents and models make live decisions.

Manual work declines as system-driven action replaces repetitive execution, freeing capacity for strategy and creativity. Actions align across channels, creating consistency customers can feel. Most importantly, ROI becomes attributable. Gains in conversion, retention, and revenue trace back to how the system operates rather than heroic effort or isolated campaigns.

The shift to modern marketing is measured by operational velocity, adaptability, and sustained business impact. When these signals appear consistently, the organization is no longer oscillating between states. It's achieving modern marketing performance.

Summary

Many marketing organizations appear active, well funded, and technologically advanced, yet still struggle to improve performance. The constraint lies in the growing mismatch between how marketing is organized and how intelligence now moves.

Most organizations operate in a state of partial transformation. The AIM OS Transformation Journey makes these patterns visible, showing how organizations move through Decline, False Progress, Stall, and toward a true Performance System. Many remain in the middle, where old and new models coexist, increasing cost and friction while failing to deliver the benefits of either. The AIM OS Organizational Health Scorecard helps leaders see where they operate today by revealing how work, decisions, activation, learning, and governance function together as a system.

Marketing's role isn't to trail execution. It's to help define where the organization must go next. Breaking out of the middle requires structural redesign rather than incremental optimization. Progress begins with change in high-impact areas. As organizations advance, the signals shift. Cycle times compress. AI moves from pilots into execution. And manual work declines. Outcomes start to become attributable to the system's behavior rather than team effort.

Modern marketing performance doesn't come from traditional roles alone, from AI agents operating independently, or from Marketing Engineers working in isolation. It emerges from their collaboration across the operating model. For CMOs, owning performance now means owning operating design. For CEOs, the implication is that structure determines whether intelligence compounds or dissipates.

Leadership Review

1. Which transformation state best describes us today: Decline, False Progress, Stall, or Performance System?

2. Which signals of Stall or False Progress pose the greatest risk to credibility and results?

3. What evidence shows that modernization is being led with authority, budget, and cadence?

4. How does our cycle speed compare to competitors building their AIM OS?

5. Are we willing to stop funding activity that sustains legacy models?

AIM OS Organizational Health Scorecard

Every marketing transformation follows a journey, moving from legacy operating models through false progress and stall toward a true performance system. The challenge is that these states often feel similar from the inside, as activity rises and investment grows without a corresponding increase in performance. The Transformation Journey map introduced earlier makes these states visible. This scorecard helps you locate where your organization operates today by translating that journey into observable signals.

This scorecard is intentionally directional. If a question triggers debate, choose the lower option and move on. Precision comes later.

For each dimension, read the four questions top to bottom and select the first statement that accurately reflects today's reality.

Operating Model Ownership

1. There is no clear owner of the marketing operating model.
2. Ownership is discussed but not documented or enforced.
3. A named owner exists, but decision rights or cadence are inconsistent.
4. A named owner has decision rights and runs a visible weekly cadence.

Observable cue: Can leaders point to one owner and one recurring operating forum without debate?

Signals and Data Readiness

1. Customer signals arrive late and require manual reconciliation.
2. Some real time signals exist, but coverage or reliability is inconsistent.
3. Priority journeys receive usable signals with occasional delays.
4. Clean signals flow in near real time to decision points.

Observable cue: Do teams wait on reports, or do data signals trigger action automatically?

Adaptive Decisioning

1. Priority decisions are manual and routed through people.
2. Rules or models exist but are applied inconsistently.
3. Some decisions are automated, but human review is still the default.
4. Priority decisions execute automatically with intentional human oversight.

Observable cue: Are marketers deciding, or is the system deciding?

Autonomous Activation

1. Execution follows calendars and manual approvals.

2. Automation exists, but only in isolated programs or channels.
3. Continuous activation runs in select areas, not end to end.
4. Content and spend deploy continuously across channels within guardrails.

Observable cue: What can launch without a decision meeting or email approval?

Experience Loop Intelligence

1. Learning is retrospective and handled manually.
2. Insights are identified but rarely drive change.
3. Learning influences some changes, but not consistently.
4. Every interaction feeds learning back into orchestration.

Observable cue: Can teams name a change made last week because the system learned?

Intelligence Governance

1. Governance is reactive and handled after issues occur.
2. Policies exist, but enforcement happens outside workflows.
3. Monitoring is in place, but visibility or escalation is incomplete.
4. Auditability, brand safety, and compliance are embedded in flow.

Observable cue: Can the system explain why a decision fired in minutes?

Talent and Roles

1. No Marketing Engineering capability exists.
2. Engineering support is borrowed or informal.
3. Dedicated marketing engineers exist, but authority or scale is limited.

4. Marketing Engineers are staffed, empowered, and operate the system.

Observable cue: Who fixes the system when it breaks, and is that role explicit?

Value Proof to the Enterprise

1. Success is measured by activity and output.
2. Isolated wins are visible, but system impact is unclear.
3. Some system-level metrics are tracked inconsistently.
4. Leadership sees compounding gains tied to system performance.

Observable cue: Can leaders connect outcomes to system behavior rather than effort?

How to Score Your Answers

For each dimension, assign values based on the statement you selected:

Statement 1 = 0 points

Statement 2 = 1 point

Statement 3 = 2 points

Statement 4 = 3 points

Add the points across all eight dimensions to calculate your total score.

- **0–6** Decline State
- **7–12** False Progress State
- **13–18** Stall State
- **19–24** Performance State

How to Read Your Result

This score reflects how your system operates today, rather than isolated wins or individual team performance. The state you land in on the Transformation Journey indicates the dominant way work, decisions, and learning move across marketing.

If the state feels behind where you expected, assume the system is revealing its tightest constraint. Most organizations operate across multiple states at once, but performance follows the weakest one. Use the result to focus on what must change next to move forward on the journey.

What to Do Once You Know Your State

The purpose of this scorecard is not self-assessment. This is about action.

If you are in Decline: Focus on one journey, not the whole function. Map signals to decisions to actions. Remove one approval gate. Prove momentum before expanding scope.

If you are in False Progress: Pause new tools and pilots. Redesign one workflow end to end around AIM OS. Name an owner. Publish weekly metrics. Depth matters more than breadth.

If you are in Stall: Consolidate activity. Fuse pilots into a single operating flow. Increase decision automation coverage. Shift leaders from coordination to system oversight.

If you are operating as a Performance System: Congratulations. Raise targets. Expand guardrails. Increase reuse. Move journey by journey. Your advantage now comes from compounding learning and cycle speed.

PART III

Running the AI Marketing Operating System

9

Running AIM OS at Full Speed

*The engineered OS layers that power
each performance capability zone*

Let's look under the hood of the AI Marketing Operating System. AIM OS creates advantage when its four performance capability zones operate as a connected system across *Adaptive Decisioning, Autonomous Activation, Experience Loop Intelligence*, and *Intelligence Governance*. The breakthrough is much more than new AI technologies. It's the ability to connect intelligence end-to-end so marketing can operate in continuous motion.

When this happens, signals inform decisions. Decisions trigger actions. Actions generate learning. Governance keeps speed sustainable. And when this flow is engineered, marketing performance compounds. But when it fragments, gains remain largely incremental.

In practice, many organizations already have pieces of this flow in intent data tools, relationship management systems, and customer analytics platforms. Decisions are embedded in models, rules, and planning processes. Actions execute through marketing automation, media platforms, and program workflows. The constraint is clearly not a technology issue. It's an integration and orchestration issue. A portfolio of martech capabilities rarely operate with shared intent or as a single operating sequence.

This chapter focuses on how that sequence runs. How signals are consolidated into decision ready inputs. How decisions are translated into specific instructions rather than recommendations. How actions are executed without waiting for manual coordination. How learning feeds back into logic while market conditions are still changing. And how governance is embedded so speed scales without risk.

The pages ahead serve as playbooks for CMOs and their teams. They show how each zone operates in practice and how to recognize whether a zone is healthy or constrained. The focus is on engineering operating flow so marketing can interpret signals and activate autonomously. This is the new how. And in this era, advantage belongs to teams that fully leverage and orchestrate their existing technology investments to deliver modern marketing performance.

Playbooks That Perform

These lessons come from work across industries, organization sizes, and technology environments. We've seen them inside teams managing complex martech portfolios and inside organizations with fewer tools but tighter operational alignment. The pattern is consistent. Essential capabilities are present, but they're rarely engineered to operate as a single system that can put intelligence in motion. When teams stop thinking in stacks and start thinking in an operating flow, a completely new level of performance is unlocked.

The examples in this chapter are drawn from real engagements, anonymized but intact in their lessons. You will see the same patterns recur across different contexts because the constraints are structural, even when organizations differ.

Before we go deeper into each zone, let's bring AIM OS back into view. In Chapter 3 we introduced it as a system built to sense, interpret, decide, allocate, activate, and learn in continuous motion. It represents the structural shift that modern marketing leaders now need to deliver growth in the AI era. In Chapter 5 we showed how existing data, martech, and AI tools are integrated to power each capability zone. At this point, it is important to see how AIM OS works through an operational sequence across the four connected zones.

Exhibit 16: **AI Marketing Operating System in Motion**

Zone	Role in the System	AI Function	Core Motion
Adaptive Decisioning	Determines priorities, allocations, and next best actions	Agents monitor signals and determine decisions; Generative AI assembles content	Turns intelligence into decisions
Autonomous Activation	Executes decisions across channels in real time	Agents push decisions to platforms; generative AI delivers variations	Turns decisions into actions
Experience Loop Intelligence	Captures outcomes and updates future decisions	Agents detect patterns; generative AI summarizes learning	Turns actions into learning
Intelligence Governance	Enforces standards, visibility, and accountability	Agents manage consent and auditability; generative AI supports explainability	Keeps performance sustainable

Together, these zones form a single operating system that becomes faster and sharper with every cycle. What makes this model urgent for CMOs is its precision and velocity. Intelligence now advances simultaneously across all four zones.

Real advantage comes from building the capability, and the confidence to operate it at speed. The challenge is making it perform across the complexity of today's data, technology, and skills. This isn't about parity or industry standard. It's about building something competitors cannot easily copy.

Engineering AIM OS is a competitive advantage
that can't be easily acquired.

As intelligence becomes embedded across marketing, execution shifts from teams to agents operating inside AIM OS. AI agents act as delegated operators. They monitor signals, carry out decisions within defined bounds, coordinate activation across channels, and feed learning back into the system. In summary, teams set intent and constraints and agents run the system at speed.

Before moving into the mechanics of each zone, one principle matters. And that is building AIM OS is a cycle of design, execution, and refinement. Pick a zone or a workflow and then engineer intent. And then run it, measure it and refine it. Each iteration teaches the organization how intelligence moves through the system. Over time, the work shifts from adding tools to tuning connections. This is the operating mindset of modern marketing.

With that foundation in place, let's step into the mechanics of each performance capability zone. The following sections outline the operating layers within each zone, including the marketing technologies typically involved. Your specific technology stack will differ. The purpose here is to illustrate the types of marketing technologies commonly represented in each zone, not to prescribe a fixed solution. Let's explore further:

Capability Zone 1: Adaptive Decisioning

Adaptive Decisioning is where modern marketing performance is won or lost. It determines how quickly an organization can convert live customer and market signals into decisions about who to engage, what to deliver, where to act, and when to move. In an AI driven market, relevance is built in hours, rather than quarters. Decision latency now becomes your operating constraint.

Historically, marketing decisions were made in planning cycles. Teams reviewed past performance, debated priorities, and committed budgets months in advance. That approach assumed stable behavior and predictable journeys. This approach no longer works. Buyers now signal intent continuously across digital, physical, and new AI environments. And this increasingly complex portfolio of data and signals quickly loses value if they aren't interpreted and acted on in real time.

Decisioning happens increasingly through AI agents responsible for evaluating live signals and executing next best actions within defined thresholds. Decisions are made in context. When allocation adjusts dynamically, the result is better judgment applied faster.

A global industrial equipment manufacturer illustrates the shift. The company struggled with long sales cycles and uneven pipeline quality. Marketing ran fixed nurture programs built around email cadences, gated content, and scheduled webinars. Activity was high, but outcomes were inconsistent. High-potential accounts stalled while lower-value prospects consumed disproportionate attention.

The breakthrough came when the organization introduced an Adaptive Decisioning layer. Their customer data platform (CDP) was unified with signals from web behavior, content, engagement, customer relationship management (CRM) activity, and third-party intent data. When buying signals crossed defined thresholds, the system determined the appropriate response and routed it directly into marketing and sales workflows. The system automatically triggered the next best action, including a personalized sales alert, tailored content sequences and dynamic ad suppression directly within existing workflows. Within ninety days, marketing contribution to pipeline increased by twenty-eight percent and average deal velocity improved by seventeen percent. More important, sales reported a visible improvement in lead quality. The performance advantage came from shortening the distance between customer signals and marketing actions.

At the operating level, Adaptive Decisioning rests on three tightly connected orchestration layers. These are:

1. **Signal Aggregation** consolidates behavioral, transactional, and contextual inputs into a unified decision view. Signals only matter when they can be tied to a specific account, decision

maker, and stage of intent. Fragmented data creates noise, but integrated signals create clarity.

Most marketing organizations already capture these inputs through web analytics, social listening, sentiment tools, CRM, marketing automation platforms (MAP), data management platforms (DMP), third-party intent providers, and conversational intelligence. Within Adaptive Decisioning, these feeds are unified around customers and monitored continuously. AI driven monitoring detects readiness or risk as the signals emerge. The objective isn't more data, but decision-ready signals prepared for immediate interpretation.

2. **Decision Modeling** evaluates signals against defined objectives and constraints. Predictive logic prioritizes opportunities based on likelihood and urgency. Rules-based orchestration applies business conditions such as account priority, promotion periods, budget limits, and brand standards. Generative capabilities shape messages and offers after a decision is made. The system determines what action is warranted, and what content is available.

   In practice, this logic extends marketing automation platforms, customer data platforms with embedded decisioning, predictive analytics tools, and marketing attribution platforms. These models are embedded directly into the decision path rather than sitting in separate reporting layers. Thresholds are tuned as models learn. Decisions are explicit, repeatable, and owned. At this layer, insight no longer stops at recommendation. Now it drives behavior.

3. **Real Time Instruction** translates decisions into relevant execution directives. Each decision specifies who to engage, what to deliver, where to act, and why. Those instructions move immediately into activation environments and adjust continuously as conditions change.

Marketing automation platforms, demand-side platforms (DSP), ad servers, search engine marketing platforms (SEM), social advertising platforms, account-based marketing platforms (ABM), sales enablement and engagement platforms, SMS and mobile messaging platforms, on-site personalization engines, and email service providers (ESP) all receive these directives simultaneously. Execution shifts from campaign schedules to responsive action. AI driven adjustment continues with guardrails set by governance, ensuring speed and reducing drift.

Adaptive Decisioning doesn't rely on a narrow slice of data or a handful of signals. It evaluates the full range of inputs available to the organization in real time, from behavioral activity and transactional history to contextual signals, intent data, and in-market actions across channels. The value comes from synchronous reasoning. Signals are assessed together, instead of sequentially, and decisions are made in context rather than in isolation. This is a capability never thought possible for marketing.

Adaptive Decisioning closes that gap. It allows intelligence to operate at a scale and speed no team could ever replicate, while freeing marketers to focus on judgment, strategy, and creative direction. Roles aren't displaced. Instead, team reasoning and decisioning are expanded.

For CMOs, Adaptive Decisioning introduces a new form of accountability by requiring ownership of how decisions are made. This includes defining which signals matter, setting thresholds that balance speed and precision, and ensuring decision logic reflects business priorities and brand standards. When decision design is unclear, execution quality cannot compensate.

Adaptive Decisioning is foundational because *every* other zone depends on it. Without timely decisions, execution slows and learning degrades. Governance becomes reactive. When decision velocity improves, the next constraint emerges immediately, and that is the ability to move those decisions into market without delay. That is the role of Autonomous Activation.

Capability Zone 2: Autonomous Activation

Autonomous Activation is the execution layer of AIM OS. It converts decisions into live market action instantly and at scale. Once Adaptive Decisioning specifies who to engage, what to deliver, where to act, and when to move. This zone carries these instructions directly into execution environments without waiting for approvals, handoffs, or manual scheduling. The result is marketing that executes as a continuous flow rather than a mosaic of queued campaigns.

In traditional marketing, execution is the constraint. Decisions may be sound, but they stall in budget allocations, program approvals, and fragmented workflows. Market opportunities fade as teams wait their turn. Autonomous Activation removes these issues by treating execution as a system behavior rather than a team bottleneck.

A global B2B software provider relied on campaign calendars and manual coordination to move from insight to activation. Each touchpoint required management intervention. Sales cycles slowed as marketing struggled to keep pace with live opportunities. The shift came when the organization embedded an Autonomous Activation layer across their demand programs. Instead of scheduling campaigns, the system executed next best actions defined by Adaptive Decisioning. A high-value prospect downloading a whitepaper received a personalized follow-up within minutes. Webinar engagement triggered sequenced media and content steps already mapped by the decision layer. Generative AI assembled brand compliant creative within approved templates, with adapted variations based on performance. After one quarter, pipeline acceleration increased nineteen percent and campaign cycle time was cut in half. Execution speed became a competitive advantage.

At the operating level, Autonomous Activation is built from several tightly coordinated capability layers that translate decisions into action. These are:

1. **Trigger Intake** receives approved decision packages and qualifying events such as form fills, demo requests, content engagement, or conversational signals. These triggers are routed into named workflows without reinterpretation. Most organizations already support this through marketing automation platforms, customer data platforms, web and product analytics platforms, chatbot and conversational AI platforms, form and lead capture tools, event and webinar platforms, and consent management platforms. The shift is using these tools as execution endpoints and not campaign schedulers.

2. **Dynamic Content Delivery** assembles messages and experiences that match context, channel, and audience. Generative AI fills structured templates with approved language, offers, and formats. Technical evaluators receive product depth. Executives see business outcomes. Site visitors experience modular personalization. Content management systems (CMS), digital asset management platforms (DAM), personalization engines, product information management systems (PIM), email template builders, dynamic creative optimization platforms (DCO) and recommendation engines already provide the infrastructure. AIM OS provides the instruction.

3. **Channel Orchestration** ensures actions deploy across email, paid media, social, messaging, and conversational environments as a single sequence. Email service providers, demand side platforms (DSP), ad servers, search engine marketing platforms, social advertising platforms, retail media networks, programmatic audio and video platforms, SMS and mobile messaging platforms, push notification services, and affiliate marketing platforms execute instructions consistently while frequency, suppression, and sequencing rules remain centralized. Channels no longer operate as independent silos. They act as coordinated endpoints.

4. **Workflow Automation** connects marketing, sales, and service so actions are logged, scored, and handed off without issues. Customer relationship management systems, sales engagement platforms, integration and middleware platforms, revenue operations platforms, customer success platforms, collaboration and project management tools, and ticketing systems ensure execution triggers the next step when team involvement is required.

5. **AI Search Activation** routes decisions into AI mediated discovery. environments. Structured data, product feeds, and content management allow offers, positioning, and differentiation to surface accurately inside generative search and large language model (LLM) interfaces. Schema and structured data tools, product feed management platforms, search engine optimization platforms, content management systems, knowledge bases and FAQ platforms, and digital experience platforms (DXP) support visibility in AI mediated discovery. As discovery shifts beyond brand controlled environments, execution must travel with it.

Together, these capabilities form the execution muscle of AIM OS. Once a decision is made, this zone ensures it reaches the market immediately, consistently, and safely. Every action is traceable and every response becomes input for learning.

For leaders, the mandate is operational. Start with a single high-frequency decision flow such as demo requests, inbound content engagement, or direct response media. Wire that decision path into core execution platforms like email, paid media, and web. Confirm that a signal can generate, assemble, and deploy an action automatically, end to end, without manual intervention. The objective again is intelligence in motion.

Next, calibrate execution. Align data models, triggers, suppression logic, and consent rules so platforms follow system instructions precisely. Most friction in this zone is self-imposed, caused by mismatched flows, incomplete templates, or unclear ownership. And as expected, this still requires creative readiness. Maintain a living asset kit mapped to top decision types, so the system never waits for production. Use generative assembly to adapt within approved boundaries.

When Autonomous Activation is working, execution stops feeling constrained. Campaigns unfold. Personalization happens in motion. And every response feeds the next decision.

Once actions are live at speed, performance depends on learning. That responsibility belongs to Experience Loop Intelligence.

What AI Agents Do in AIM OS

AI agents are the operators inside AIM OS. They carry out work that requires speed, consistency, and continuous attention, operating within boundaries set by leaders and Marketing Engineering.

In practice, agents monitor signals as they emerge, evaluating changes in customer behavior, context, and performance across channels. They apply decision logic in real time, recommending or executing next actions within defined thresholds rather than waiting for manual review.

Agents also coordinate activation. They translate decisions into action across platforms, channels, and experiences so execution stays aligned as conditions change. Timing, sequencing, and handoffs remain consistent even as volume and complexity increase.

As outcomes occur, agents observe responses and feed learning back into the system. They compare what happened to what was expected, refine future decisions, and improve performance with each cycle. Learning becomes continuous rather than episodic.

Governance is embedded in this flow. Agents enforce consent, apply brand and policy rules, and maintain auditability as part of normal operation. Speed and control move together.

AI agents don't replace marketing teams. They extend them. Teams define intent, priorities, and constraints. Marketing Engineering designs the operating model and oversight. Agents run the system at speed, allowing marketing to operate continuously in an AI driven environment.

Capability Zone 3: Experience Loop Intelligence

Experience Loop Intelligence is the learning layer of AIM OS. It captures how customers respond in real time, feeds that intelligence back into decisioning, and adjusts the next interaction as conditions change. The objective is simple in that every interaction improves the next one.

Traditional marketing teams learn too late. Performance is reviewed weeks or months after execution, long after conditions have shifted. Insights arrive as reports rather than instructions. By the time teams react, the opportunity has passed. Learning is accelerated by agents responsible for observing outcomes, comparing responses, and adjusting the next cycle automatically. This enables a level of agility never considered possible in marketing.

A direct-to-consumer fitness equipment brand demonstrates the value created. Product launches drove strong initial engagement, but follow-up performance dropped quickly. The issue wasn't creative quality or channel mix, it was response latency. The organization activated Experience Loop Intelligence to capture and interpret response signals as they occurred. AI driven analysis monitored site behavior, app usage, cart activity, and micro-signals such as scroll depth and video engagement. Generative capabilities refreshed creative in flight, adjusting imagery, calls to action, and offers based on response patterns. Learning flowed directly back into decisioning, where budgets and targeting shifted toward what worked and away from what didn't. After two quarters of system refinement, customer acquisition costs fell twenty-two percent and sales increased fifteen percent without additional budgets. Performance improved because learning kept pace with behavior.

Experience Loop Intelligence is built from three tightly connected capabilities layers that close the learning gap. They are:

1. **Signal Capture** instruments every meaningful interaction across digital, physical, and AI mediated environments. Web and product analytics platforms record on-site behavior. Marketing automation platforms and customer data platforms log engagement. E-commerce platforms, point-of-sale systems, and subscription management platforms track transactions. Social listening tools, call tracking and recording platforms, contact center platforms, and conversational AI platforms capture dialogue. Survey and feedback tools collect direct customer input. As generative search and large language model (LLM) interfaces mediate discovery, questions and responses become additional learning signals. The objective is coverage not volume. Signals must be timely, attributable, and tied to real journeys.

2. **Response Analysis** determines what is working and why. Marketing attribution platforms measure channel and touchpoint contribution. Customer intelligence platforms detect fatigue, mismatch, or decay. A/B testing and experimentation platforms validate what resonates. Marketing mix modeling (MMM) tools evaluate spend effectiveness. Generative AI summarizes patterns and proposes variations across formats, including copy, imagery, video, and interactive modules. Analysis is evaluated against business outcomes, instead of only engagement. Learning becomes directional, instead of descriptive.

3. **Feedback Integration** routes learning directly back into decisioning and activation. Bids adjust. Budgets reallocate. Experiences adapt.

Marketing automation platforms, demand-side platforms, search engine marketing platforms, social advertising platforms, personalization engines, and dynamic creative optimization platforms receive updated instructions automatically. Customer relationship management systems and customer data platforms record provenance, so changes remain traceable. Consent, privacy, and explainability controls ensure learning remains compliant as AIM OS accelerates.

When these capabilities operate together, learning compounds. Marketing no longer waits to understand what worked. It adapts while it is happening. Each response sharpens the next decision. Each cycle reduces waste and increases relevance.

For leaders, the mandate is to design for flow rather than reporting. Choose an experience that runs frequently enough to reveal patterns like onboarding, nurture, or a core content series. Ensure feedback from that experience flows directly into decision logic. And review performance weekly through the lens of outcomes, rather than a status of activity.

Common failures are structural. Vanity metrics inflate confidence without improving relevance. Over-targeting erodes trust when pacing, context, or consent are ignored. The remedies are design choices versus tool changes. Define which signals matter, where learning must return, and how often adjustments occur.

When Experience Loop Intelligence is active, campaigns evolve, offers adjust, and budgets move. Marketing stops reacting to outcomes and starts shaping them continuously in motion. This is how relevance compounds.

With learning in motion, the final constraint becomes control. Speed without trust breaks systems. That responsibility belongs to Intelligence Governance.

Capability Zone 4: Intelligence Governance

Intelligence Governance is the control layer of AIM OS. It ensures that speed, automation, and intelligence operate within accountable boundaries. As AI driven decisioning and activation accelerate, unmanaged risk becomes a performance threat. Without transparency, intelligence erodes. And without trust, program velocity slows. Governance enables sustainable speed by embedding responsibility, traceability, and explainability into every signal, decision, and action.

It's important to note that this isn't back-office hygiene. It's an equal dimension of AIM OS. In a market where every brand increasingly claims intelligence, the leaders who win are the ones who can prove it. This means it's responsible, explainable, and aligned with business intent. Regulators, customers, and boards now evaluate performance and trust together.

A retailer illustrates the point. Campaign execution slowed under growing regulatory scrutiny. Consent gaps and data quality issues triggered escalations. Approval cycles stretched. What looked like a compliance problem was, in fact, an operating failure. Speed and trust had broken together. By embedding automated consent enforcement and live governance monitoring directly into execution workflows, the retailer reduced approval time by forty percent, lowered regulatory exposure, and restored confidence. Governance didn't slow performance, it unlocked it.

At the operating level, Intelligence Governance is built from several interlocking capability layers that keep AIM OS visible, compliant, and controllable at speed.

1. **Policy and Control Frameworks** define what responsible and explainable intelligence means for the organization. Policy management platforms, AI governance platforms, model registries, explainability and AI transparency tools, documentation and knowledge management systems, digital asset management plat-

forms for approved content governance, and workflow approval tools establish shared rules that the system enforces automatically.

2. **Data and Consent Management** ensures that all inputs used by AIM OS are permissioned, current, and compliant. Customer data platforms with embedded consent enforcement, consent management platforms, privacy management platforms, data quality and observability tools, data catalogs, master data management platforms, identity resolution platforms, and preference management centers govern how data can be used before it enters decisioning or activation.

3. **Model Oversight and Auditability** makes algorithmic behavior observable and reconstructable. Machine learning governance platforms track model versions, monitor drift, and log decisions. Telemetry dashboards provide visibility into how recommendations were generated and why they changed over time.

4. **Process Guardrails** accelerate flow rather than block it. Marketing resource management platforms, workflow and approval automation tools, brand safety platforms, ad verification and fraud detection tools, legal and regulatory review systems and creative compliance platforms ensure that execution moves quickly while staying within defined boundaries. Governance shifts from exception handling to design.

Agents enforce consent, monitor compliance, and maintain auditability as part of normal execution. When these capabilities operate together, governance truly becomes a performance multiplier. Decisions move faster because they are trusted. Automation scales because it produces traceable signals. Learning accelerates because it is explainable.

For leaders, the mandate is clarity. Legal, risk, and compliance teams must understand what decisions the system is making, why they occur, and where data originates. This shared visibility builds confidence in automation and removes friction from execution. Strong governance doesn't constrain momentum, it sustains it. It gives executives visibility, regulators assurance, and customers confidence that intelligence is working in their interest.

With governance in place, AIM OS can run faster without breaking trust. Together, the four zones form an operating system that senses, decides, activates, learns, and governs as one continuous flow.

From Operating Model to Operating Discipline

Designing AIM OS is only the first step. Running it is the real work. Across organizations, the most common failure isn't technology readiness or executive intent. It's execution ownership. Teams invest in data platforms, automation, and AI capabilities, but no one owns how those pieces operate together. The connections remain fragile, manual, or improvised.

An operating system without discipline simply won't perform.

Across nearly every engagement, we see the same gaps appear. Organizations possess many of the essential capabilities required to run AIM OS, but they aren't engineered to operate as one system. Intelligence exists, but it doesn't flow. Decisions are made, but they don't execute efficiently. And learning happens, but it doesn't compound.

This is why actual AIM OS performance remains rare. If operating intelligence at speed were easy, it simply wouldn't be an advantage. The differentiator isn't technology access. It's the discipline required to connect, run, and continuously tune the system.

This is where many AIM OS efforts stall. CMOs sponsor the vision. Teams deploy some new tools. But without a discipline responsible for designing, connecting, and tuning the system end to end, the promise and potential of intelligence is never realized. Running AIM OS requires a different capability.

Marketing Engineering: The Operating Discipline

Marketing Engineering is the discipline that designs how AIM OS operates. The CMO defines strategy and priorities. Marketing Engineers design and manage the operating model that allows those priorities to run continuously as AI accelerates signals, decisions, and execution.

Marketing Engineers determine how data, technology, and workflows connect so marketing can sense, decide, and act as one system. They shape how decisions are made, how activation stays coordinated, and how learning improves performance over time. Their work shapes how marketing behaves and intelligence flows across the four zones of AIM OS.

Marketing Engineers are fluent in traditional marketing and modern marketing technology. They bridge marketing intent and technical execution, partnering closely with teams to translate marketing objectives into system behavior the organization can rely on.

In practice, this means designing the data foundations that keep signals consistent and trusted as they move through the system. It means integrating marketing technology so platforms share context and decisions carry forward. Marketing Engineers design the operating model and define the logic, boundaries, and oversight within which AI agents operate, so execution can be safely delegated at scale.

As AI becomes central to marketing performance, Marketing Engineering becomes essential. It turns strategy into a working system. It allows AIM OS to operate with speed and reliability. And it gives CMOs confidence that marketing can perform as customer behavior continues to evolve.

This is an engineering mindset applied to marketing. Build the model, run it in market, learn from performance, and continuously improve how the system behaves as conditions change.

For most organizations, building this capability entirely in house takes time. As a result, many CMOs establish Marketing Engineering as a hybrid capability, combining internal leadership with experienced external support. This approach compresses time to value, reduces risk, and allows the operating model to take shape incrementally, and with confidence as performance improves. This is why Marketing Engineering has become an essential operating discipline for modern growth.

Summary

Modern marketing performance is driven by how well intelligence moves through the organization. The four capability zones of AIM OS describe what modern marketing must be able to do: sense changing conditions, make decisions in motion, activate coherently across channels, learn continuously, and operate with governance built in. Together, they define the behavior of a marketing organization built for an AI driven environment.

What determines whether those zones actually work is the enabling layers beneath them. Each zone depends on engineered foundations that allow intelligence to flow smoothly. These include connected and trusted data, decision logic that operates in real time, activation systems that coordinate across channels, learning mechanisms that improve performance with every cycle, and governance structures that scale safely. These layers may sit out of view, yet they decide whether the operating model holds together as speed and complexity increase.

This is why AIM OS cannot be assembled through tools alone. The zones require intentional design across data, technology, workflows, and controls so they function as one system rather than a set of independent capabilities. Marketing Engineering is the discipline that designs and maintains these layers. It translates strategy into system behavior and ensures the operating model runs reliably as conditions change.

This chapter shows how modern marketing performance is built. The competitive advantage comes from engineering the operating model that allows intelligence to move continuously from signal to decision to action to learning. That is the foundation for sustained performance in the AI era.

Leadership Review

1. How does intelligence move through our marketing organization today?

2. How quickly do signals become decisions in Adaptive Decisioning?

3. Where do handoffs slow execution in Autonomous Activation?

4. How much does each learning cycle improve the next in Experience Loop Intelligence?

5. Does Intelligence Governance preserve trust without sacrificing speed?

10

The Metrics That Drive Performance Now

Applying the system metrics that
power modern marketing performance

A CMO asked us to review his quarterly marketing update before presenting it to the leadership team. At first glance, the numbers looked solid. But one chart was challenging. Qualified marketing leads had dropped to 24% of total pipeline contribution for the month, less than half their run rate baseline of 37%.

When we asked why, the Marketing Operations leader revealed that Sales Ops had added a new field validation rule in the CRM platform. But no one had informed marketing. As a result, leads from ABM campaigns and virtual events were silently being rejected in reports before they ever reached sales. By the time the report was prepared, dozens of opportunities had vanished, and the damage was done.

This team was already using autonomous campaign tools to launch and optimize programs. But the reporting layer hadn't kept pace with the performance layer. The system could execute in real time, but leadership only saw problems weeks later, after revenue opportunities had passed.

I asked the CMO what the meeting would look like if he had seen the problem in real time rather than after the quarter closed. He acknowledged that marketing was now running faster than its measurement.

That was progress, but the scoreboard still looked backward. It reported outcomes after value was lost, not signals while decisions could still change the result. A scoreboard only matters if it guides decisions along the customer journey and adjusts strategy in motion. Yet most marketing scoreboards still focus on the past. They fill operating decks but rarely show performance patterns while they still matter.

In the AI marketing era, what's missing is operating scorecard alignment. Outcomes, operating rhythm, and live signals cannot sit in isolation. They must come together in a single, actionable view.

AI agents learn, adapt, and compound, but that compounding only works if leaders can see the results and the condition of the execution layer producing them. This is the gap the *Modern Marketing Dashboard* closes. It brings together three forms of performance visibility in one place: outcomes achieved, operating indicators driving them, and forward trends that show where momentum is heading.

With this view, the CMO can run marketing in real time. They see what's working, why it is working, and where it is heading early enough to shape the outcome. In an AI driven market where advantage is measured in hours and days, marketing shifts from reporting the past to steering performance as it happens.

Three Views For Modern Marketing Leadership

With the AI Marketing Operating System, marketing now operates at a cadence and speed once unimaginable. It is the difference between driving by the rearview mirror and navigating by GPS. Performance depends on how fast, how precisely, and how reliably the system senses, decides, and acts. Today's CMO is accountable for an integrated operating model that performs under real time conditions.

This changes the nature of measurement. It's no longer enough to report outcomes alone. CMOs must also see how the operating rhythm producing those outcomes is performing and intervene before results slip. Without operational and forward-looking visibility, even strong creative or aggressive demand plans can be undermined by delays and breakdowns that surface only after value is lost.

Other environments have operated this way for decades. In manufacturing, leaders track line speed, defect rates, and uptime alongside output. In competitive racing, teams monitor temperature, tire wear, and fuel consumption to adjust mid-race. In logistics, companies watch port delays and fleet utilization to keep supply chains moving. Performance depends on seeing the system, not just the result.

No high performance field runs on outcomes alone. Every profession of consequence manages the system that drives its results. Pilots, engineers, traders, and athletes all know that advantage comes from sensing, deciding, and correcting in motion. Marketing now joins these groups. With AIM OS, CMOs can lead with the same control and precision once reserved for mission-critical operations.

The **Performance View** shows what happened. The **Operating View** explains how it happened. The **Telemetry View** reveals what is happening now and where momentum is heading. Together, they turn marketing's dashboard from a static report into a live, modern management instrument.

Exhibit 17: **The Modern Marketing Dashboard**

One system. Three lenses. Unified visibility.

Metric Views	Purpose	Examples
Performance View	The scoreboard of what happened	• Outcomes: engagement, demand response, revenue • Establishes credibility • Risks: invites debate without context
Operating View	The diagnostics of how it happened	• System health • Metrics: throughput, cycle times, first pass effect, WIP, uptime • Reveals scalability
Telemetry View	The live view of what's happening now and where it's headed	• Forward-looking signals • Metrics: OME, value leakage, conversion momentum, bottlenecks, model drift detection

In every high-performance field, leaders demand all three. In marketing, AIM OS makes this possible by fusing data, automation, and intelligence into one living system. For CMOs and CEOs, the implications are threefold. Outcomes build credibility. Operating rhythm enables scale. Telemetry sustains competitiveness. When combined, these views show how marketing is actively engineering performance.

The New Modern Marketing Dashboard

If the three views form the foundation of the new framework, the next question is what to measure. A dashboard is only as useful as the gauges you include. Too many metrics create noise. Too few hide early warnings and eliminate the chance to act in time. For CMOs, striking the right balance is challenging.

The *Performance View* is well understood. Program engagement, revenue impact, pipeline contribution, and conversion rates are already standard. We won't revisit those here. What's been missing are the *Operating View* and *Telemetry View*. These layers reveal whether performance outcomes are sustainable, and whether intervention is possible before results slip.

Across mature performance disciplines, the critical measures have been codified over decades. Manufacturing leaders track throughput, defect rate, and downtime. Racing teams monitor live data to hold competitive pace. Logistics networks measure delivery precision to anticipate constraints before they occur.

Marketing can now operate with the same discipline. AIM OS makes it practical to monitor a focused set of Operating and Telemetry measures that show how performance moves through the system in real time. For the CMO, this becomes the command layer. This is now a live view that replaces lagging reports with actionable intelligence.

For CMOs, these metrics bring clarity to what has always felt complex. They show whether marketing is keeping pace with demand, whether programs convert as expected, and whether the system is learning fast enough to stay ahead. They replace endless reporting with a set of key indicators that show where to lean in and where to steady execution. For CEOs, they provide confidence that growth is scalable and that marketing can expand impact without inflating cost or losing speed in the market.

In the sections that follow, we introduce ten new measures to complement existing outcome metrics. Operating Metrics show how well AIM OS is running today and whether it can sustain performance at scale. Telemetry Metrics provide live visibility into where performance is heading, enabling teams and AI agents to adjust before results slip. Together, these ten measures move marketing from reporting the past to engineering performance in motion.

Performance Metrics (Your Existing Measures)

Most CMOs already manage these metrics extremely well. Brand health, campaign engagement, conversion rates, and revenue impact have long defined marketing's contribution to growth. They remain essential for proving results, but they only show the finish line. What they don't reveal is *how* performance is being created or *where* it might be at risk.

That's where these two new sets of measures come into play.

Operating Metrics (New)

The Operating Metrics show whether the system producing results is running at full speed. They measure performance against goals and reveal how efficiently AIM OS turns signals into action, indicating whether the operating rhythm can sustain performance at scale. These metrics are:

Throughput: Velocity of Value Creation
Throughput measures how much meaningful marketing output your system delivers in each period. This isn't campaign volume. It reflects qualified opportunities, accepted offers, and active engagements. Think of it as the units per hour gauge on a factory floor and proof that the system is delivering against growth goals. If throughput is sluggish despite strong resources, the system is the constraint and it's underperforming.

Cycle Time: Speed from Signal to Market Action
Cycle Time measures how long it takes for a customer signal to become a live, relevant market response. A search query, a form fill, or a cart abandonment should trigger action within minutes, not days. In racing, pit stops are timed in seconds because every delay costs position. Marketing is no different. If your system reacts slower than your competitor's, you are not on track to win.

Yield (First-Pass Effectiveness): Getting it Right the First Time
Yield measures the percentage of marketing actions that achieve the intended outcome on the first attempt. Accepted offers, converting ABM plays, and content reaching the right audience without rework all signal high yield. In manufacturing, first-pass yield reduces defects and waste. In marketing, it reflects precise targeting, efficient spend, and customer trust. A system with high yield consistently delivers value without unnecessary cycles.

Work in Progress (WIP): Active Marketing Load
Work in Progress reflects the total number of campaigns, programs, and journeys active at any time. Too much WIP clogs the system, stretches teams, and slows cycle time. Too little means underused capacity. Experienced operators manage WIP carefully to keep production flowing at peak efficiency. In marketing, balanced WIP sustains momentum without burnout or execution risk.

Uptime: Reliability of the Automation Layer
Uptime measures how consistently critical platforms such as CDPs, automation engines, and AI orchestration layers are available and performing as designed. In logistics, downtime delays deliveries. In marketing, downtime means missed signals, untriggered offers, and lost revenue. High uptime is the foundation of scale, and its absence is noticed only when the martech fails.

Telemetry Metrics (New)

While Operating Metrics show how well AIM OS is performing today, they only tell part of the story. Modern marketing also needs visibility into what is happening now and what is likely to happen next. Telemetry Metrics provide this forward view. They function as an early warning and opportunity system, revealing where momentum is building, where friction is emerging, and where performance can be tuned before results slip. Think of them as your weather or fitness app for marketing. These metrics are:

Overall Marketing Effectiveness (OME): Composite Performance Signal
OME combines multiple indicators into a single top level view of marketing's health. It functions like dashboard warning lights. Green signals stability. Yellow signals attention. Red signals risk. When leaders cannot track every metric, OME provides a fast, credible read on whether marketing is performing as expected. It can blend measures such as Cycle Time, Yield, Uptime, Value Leakage, and Drift, and be recalibrated quarterly as conditions change.

Value Leakage: Where Revenue Escapes
Value Leakage identifies where potential growth is slipping away before it can be realized. Leads that stall, offers that fail to trigger, or customers who abandon because response was too slow all represent leakage. In supply chains, leakage means goods lost before delivery. In marketing, it means potential revenue disappearing. Tracking leakage prevents loss and often recovers growth faster and more efficiently than chasing new demand.

Conversion Momentum: Trajectory of Market Response
Conversion Momentum shows whether conversion rates are rising, holding steady, or declining across active programs. It does more than report outcomes. It reveals the direction of performance in real time. Like a stock chart, the slope matters as much as the number. Momentum tells leaders when to double down, when to adjust, and when to intervene before results deteriorate.

Resource Efficiency: Cost per Unit of Value
Resource Efficiency measures the cost to generate a defined unit of marketing value, such as cost per qualified lead or cost per incremental revenue dollar. Manufacturing has long relied on cost per unit to improve efficiency and profitability. Marketing can now apply the

same discipline. Tracking this metric shows how effectively investment turns into impact and where budgets can be dynamically redeployed for higher return.

Bottleneck Alerts: Early Signals of Constraint
Bottleneck Alerts use AI to surface friction before it slows performance. Automation queues backing up, approvals stalling, or content pipelines running dry often appear after the damage is done. Bottleneck Alerts act like traffic ahead warnings, giving teams time to clear constraints (rerouting) before weeks of opportunities are lost.

Model Drift Detection: Staying Aligned with Behavior
Drift Detection continuously checks whether models are still predicting accurately and recalibrates them before errors compound. It is like adjusting navigation while in motion rather than waiting until you are off course. Without it, performance degrades gradually and goes unnoticed until results decline.

Why These Ten New Metrics Matter

These ten metrics move marketing beyond after-the-fact reporting. They show what is working, what is slowing down, and what needs attention before results are lost. Each answers a simple, high-value leadership question about the AIM OS: how fast, how precise, how efficient, and how reliable it is.

In the end, the goal is not just to track results, but to understand and steer the system that creates them.

Exhibit 18: **The Modern Marketing Metrics**

Metrics	Definition	Why It Matters
Operating View (New)		
Throughput	Velocity of marketing value creation (qualified opportunities, offers accepted, active engagements).	Shows if the system is producing at the pace growth demands.
Cycle Time	Time from customer signal to market action.	Faster response means higher chance of winning the moment.
Yield / First Pass Effectiveness	Percentage of marketing actions that succeed on the first attempt.	Indicates precision targeting and efficient spend.
Work in Progress (WIP)	Active campaigns and journeys in motion.	Prevents overload or underuse and keeps the system balanced.
Uptime	Reliability of automation and orchestration layers.	Downtime means missed signals, delayed actions, and lost revenue.
Telemetry View (New)		
Overall Marketing Effectiveness	Composite score of system performance.	Quick warning light for executive visibility.
Value Leakage	Points where potential revenue slips out (untriggered offers, lost leads).	Identifies preventable loss and recoverable growth.
Resource Efficiency	Cost per unit of marketing value.	Proves ROI and shows where investment can stretch further.
Bottleneck Alerts	AI detection of emerging constraints.	Helps teams fix issues before they slow growth.
Model Drift Detection	Tracks AI accuracy as customer behavior shifts.	Keeps personalization and decisioning aligned with reality.

From Reporting to Real Time Control

Across the scenarios that follow, the central theme is that the CMO is no longer explaining results after the fact or defending marketing's value. They are managing performance in real time.

A Modern Marketing Dashboard turns marketing from a function that reports outcomes into one that actively steers them. Leaders see how the system behaves, where it adapts, and how it protects growth under pressure. This is a different kind of control. Less about justifying marketing's value and more about demonstrating how the discipline performs with precision. This is what modern marketing leadership looks like.

What follows shows this shift in action. Each scenario illustrates how modern marketing leadership uses operating insight and telemetry to detect issues early, intervene decisively, and keep performance on track.

Scenario 1: Identifying Execution Bottlenecks

It's Monday morning. The Marketing Leadership Team gathers for a round of status updates. This time however, the Modern Marketing Dashboard is on the screen. The CMO scans the Operating View and spots a warning. Cycle Time has jumped from two hours to nearly twelve over the past three days. Customer signals from a profitable segment are now taking days to trigger a response. In a competitive market, this really matters.

In the past, this delay would have gone unnoticed until the campaign recap report weeks later, long after the opportunities were lost. This time, the Telemetry View reveals the cause immediately. The creative production queue is jammed. New brand design templates for priority segments are still in production, creating friction across multiple programs already in flight.

The response is swift. Marketing Engineers deploy AI generated creative variants for interim use. Approvals are reprioritized for high-value accounts. Sales receives real time alerts so they can engage while the backlog clears. By midweek, the updated campaign playbook is live. Cycle Time drops below two hours and the system regains rhythm.

At the executive meeting two days later, the CMO walks through the sequence:

Signal detected. Bottleneck identified. Fix deployed. Conversion lift confirmed.

Each leader hears what matters most. The CFO sees efficiency restored. The COO sees operational control. The CEO sees adaptability and speed. This is the three views working together. Performance shows the result. The Operating View explains why. Telemetry shows where it is heading while there is still time to act.

Question: When your package delivery is delayed, you get an alert. Then why shouldn't you know when an important program is at risk?

Scenario 2: Recalibrating Personalization Drift

It's Thursday afternoon. The Telemetry View is streaming live signals across active campaigns when Model Drift Detection flags yellow. The AI engine that powers personalization shows reduced accuracy in recommending next best offers. Performance has not dropped yet, but left unchecked, sales will fall within days.

Previously, this would have surfaced weeks later through declining traffic and click-through rates. This time, the system sees it early. Telemetry traces the shift to a competitor's new pricing promotion that has changed customer behavior. The model is still working, but it is now misaligned with the market.

Product Marketing and Marketing Engineers act immediately. They retrain the model with updated competitive context, simulate new offer paths, and test variants before release. Before the weekend, the recalibrated model is live. Accuracy returns to baseline. Response rates stabilize. Early sales feedback confirms that relevance is restored.

At Monday's executive review, the CMO presents the sequence:

Drift detected. Competitive action identified. Model recalibrated. Revenue protected.

Instead of excuses weeks later, the CEO now sees revenue resilience. The CFO sees risk avoided before it became loss. The COO sees a system adapting at market speed. This is the role of Telemetry: detecting change early, diagnosing why, and enabling correction before customers notice.

Question: You know what's trending and what your kids will want to buy next. Then why should you let competitive actions that change customer preferences go unchecked?

Scenario 3: Closing Leaks Before They Drain Growth

It's mid-quarter. The Operating View shows Value Leakage climbing. Nearly 18% of qualified leads are stalling in the handoff between marketing automation and the CRM platform. Engagement is strong. Content download registrations are high. Yet many leads are disappearing before sales ever sees them.

Historically, this gap would surface only at quarter close. Marketing would claim demand. Sales would dispute contribution. Revenue would slowly slip away while teams argued.

This time, the system flags the issue in real time. Telemetry shows the leakage began immediately after a recent integration update. The Operating View traces the root cause further. A new field validation rule is silently rejecting leads through a misconfigured API.

Marketing Engineers correct the workflow, backfill missing leads into the CRM, and add an AI checkpoint to flag similar anomalies going forward. Within 48 hours, leakage rate drops below 5%. Dozens of qualified opportunities are recovered and routed to sales.

At the executive business review, the CMO doesn't just claim "the pipeline is up", while sales questions it. Instead, they walk through the sequence: Leak is detected. Integration fault identified. Workflow corrected. Pipeline protected. Leadership sees a system that can detect loss, repair flow, and prevent recurrence. This is resilience in action and a direct source of compounding growth.

Question: When your bank detects a new computer sign-on, you get notified immediately. Then why should it be any different when a major revenue process changes?

The Unified Modern Marketing Dashboard

The Modern Marketing Dashboard moves marketing beyond static reporting. It unifies three lenses so leaders can see outcomes, understand the AIM OS that produced them, and anticipate where performance is heading.

Performance View: The scoreboard of what happened

This is the familiar readout of results showing how customers engaged, how demand responded, and how much revenue or pipeline was influenced. It establishes credibility but also invites interpretation. Without context, leaders debate what the numbers mean instead of deciding what to do next. This keeps the CMO in an endless cycle of explaining, defending, and debating.

Operating View: The diagnostics of how it happened

The Operating View reveals the health of the system behind the results. It measures throughput, precision, and efficiency in converting customer signals into market action. This view shows whether marketing can sustain performance at speed and scale and where friction remains in operations. This view gives the CMO the ability to optimize resources, refine systems, and scale impact with confidence.

Telemetry View: What is happening now and where performance is heading

The Telemetry View provides live, forward-looking visibility. It surfaces shifts in momentum, emerging constraints, and early warning signals before outcomes change. This allows leaders to adjust in motion and avoid surprises that erode performance.

Used together, these three views turn dashboards into a CMO control instrument. Performance builds marketing credibility. Operating enables scale. Telemetry protects competitive advantage. The result is a system leaders can steer confidently, continuously, and at market speed. In modern marketing leadership, success is measured by how the system performs, not just how the numbers look.

Modern Marketing Performance Cadence

A dashboard has no value if it only reports what happened. Its purpose is to inform the decisions that set budgets, shape priorities, and determine growth paths. The Modern Marketing Dashboard becomes meaningful only when it is embedded in the leadership cadence, not reviewed after the fact.

For the CMO, this requires more than monitoring metrics. The three views of the dashboard function as the schematics of the AI Marketing Operating System. They reveal where throughput is strong, where value is leaking, and where predictive models require recalibration. Most importantly, they translate system health into terms that Boards understand, like financial efficiency, growth velocity, and operational resilience.

Once these metrics are in place, leaders can establish a consistent operating rhythm. The dashboard becomes a standing agenda, defining what the CMO and executive team review regularly, how decisions are made, and how performance is steered over time. Here is an example of your cadence:

Exhibit 19: **Modern Marketing Dashboard Reviews**

Operating Rhythm	Focus Metrics
Daily	Cycle Time, Value Leakage, Uptime, Bottleneck Alerts
Weekly	Throughput, Yield, Resource Efficiency, Conversion Momentum, Drift Detection
Monthly	OME recalibration, WIP rebalancing
Quarterly	Target ranges and weight tuning for CEO alignment

This is a game changer for marketing because it resets the leadership conversation. Instead of defending activity or reinterpreting past results, the CMO shows how the system is designed, tuned, and scaled. Marketing moves beyond reporting and becomes an actively managed growth system, governed with the same discipline as any mission-critical enterprise function

Summary

Modern marketing doesn't fail because of a lack of activity or investment. It fails because leaders cannot see how performance is being created while it still matters. Traditional marketing scoreboards are no longer sufficient in an AI driven environment where advantage is measured in hours, not quarters.

As marketing systems accelerate through agents and autonomous execution, measurement has lagged behind. Most CMOs can see outcomes, but not the operating health or forward signals that determine whether those outcomes are sustainable. The result is a dangerous gap where marketing can execute at AI speed while leadership only discovers problems after value is lost.

The Modern Marketing Dashboard is the control layer of the AI Marketing Operating System. It unifies three essential views of performance. The Performance View shows outcomes and establishes credibility. The Operating View reveals how efficiently the system converts signals into action. The Telemetry View provides live, forward-looking signals that show where performance is heading before results change. Together, these views move marketing from reporting the past to steering performance in motion.

System-level metrics replace legacy reporting with operational clarity. Leaders can now detect issues early, correct course quickly, and scale what works without multiplying cost. Measurement becomes a management discipline, not a retrospective exercise.

The result is a fundamental shift in leadership. Marketing stops defending periodic results and starts engineering a growth system that keeps compounding. That is the new measure of the modern CMO.

Leadership Review

1. Are we measuring performance only after the fact, or while it is in motion?

2. Which early signals allow us to adjust before results slip?

3. Do our metrics reveal system health, or only outcomes?

4. Where else in the enterprise are telemetry and operating metrics already shaping decisions?

5. Are we treating measurement as reporting, or as a core operating discipline?

11

The Double Edge of Data

Shift marketing data from overwhelming to operating advantage

At a recent marketing conference, I ran into a former colleague as the crowd spilled out from an AI panel. She laughed and shook her head. "If I hear one more person say data is the foundation of AI, I'm going to lose it." Then she paused and smiled. "Well, we've been talking about better data for a decade. Feels like now we have to prove it."

She was right. In the AI era, there's no avoiding it. Models and agents depend on data to learn, decide, and act. Modern marketing systems are now defined by the quality, structure, and reliability of the signals they consume. Marketing data can no longer be background noise. It must be an engineered foundation.

The AI Marketing Operating System (AIM OS) makes this reality practical. Each of its four zones depends on data integrity to function:

- **Adaptive Decisioning** works only when signals are reliable.
- **Autonomous Activation** scales only when context is complete.
- **Experience Loop Intelligence** improves when feedback is captured.
- **Intelligence Governance** protects only when consent and controls are clear.

Data is the essence of modern marketing performance. When engineered well, it enables precision, personalization, and speed. And when unmanaged, it multiplies noise, mistrust, and risk faster than teams can contain.

This is the double edged reality. More data will sharpen AIM OS performance, but without proper data engineering it can increase exposure.

This chapter isn't about collecting more data, building larger warehouses, or adding new platforms. Marketing already produces an abundance of data, and the industry is far along on these tactical fronts (see Exhibit 22). The work now is different. It's about running marketing data as an engineered discipline inside AIM OS, designed to power intelligence in motion.

This is where AI becomes essential. AI unlocks the foresight embedded across the entire data landscape, revealing patterns, dependencies, and opportunities no team could surface manually. The focus is what matters now to activate that advantage. Using data with specific intent rather than broad accumulation. Shaping activation decisions as they happen instead of after the fact. Designing hygiene and governance so trust scales with speed. And most important is preparing for agent ecosystems that will fundamentally change how marketing performs.

The goal is a mindset shift. Marketing data isn't a backlog to be cleaned up someday or a set of silos to be stitched together over time. It is the operating fabric of an intelligence system that senses, decides, and acts continuously. Treated this way, data compounds advantage. But treated otherwise, even the most advanced AI remains underutilized.

The Marketing Data Landscape

Data defines the landscape where modern marketing performs. It moves through every signal, channel, and decision, shaping how brands sense opportunity and respond to it. CMOs live inside this environment every day, balancing intuition with instrumentation, creativity with computation. The challenge is no longer awareness but clarity. It's seeing how these forces connect and influence performance in motion. When leaders can read that landscape in full, data becomes less of a resource to manage and more of a field of intelligence to command.

The first dimension is *customer-centric versus performance-centric*. On one side are datasets that describe real customers and clients: identities, profiles, behaviors, and engagements. On the other are datasets that describe the structures around them: revenue flows, attribution paths, media spend, and compliance frameworks that keep the system running. Both sides matter. One explains people, the other explains performance.

The second dimension is *signals versus enablers*. Some data captures what customers do, like digital footprints and actions that reveal demand in motion. Other data supports and explains those actions. It helps marketers measure effectiveness, allocate budgets, maintain governance, and prove value to the business. Signals let the AIM OS sense, while enablers let it function. And both must move together for intelligence to emerge.

Exhibit 20: The Marketing Data Landscape

	Customer-Centric	Performance-Centric
Front-end signals	• Customer identity and profiles • Behavioral and web data • App and product usage data • Campaign and engagement data • Lead and account data	• Purchase and transaction data • Attribution and journey maps • Search and SEO data • Historical performance data
Back-end enablers	• Social and community data • Advertising and media spend • Content performance metrics • Customer experience data	• Third-party and intent data • Revenue and pipeline data • Consent and privacy data • Compliance/Regulatory data

This model shows that marketing data isn't a single stream to collect or a lake to fill. It functions as an operating system of interdependent categories that create value only when managed together. CEOs can use this lens to see where capital compounds, where risk concentrates, and where performance gains are most likely to emerge. CMOs can surface hidden friction like silos that strand data, hygiene gaps that distort signals, and governance blind spots that weaken trust.

Above all, it reinforces the central theme of this chapter. Marketing data is a system discipline not a storage problem. When data is engineered with intent, it becomes the connective foundation of AIM OS, turning signals into action and action into sustained advantage.

The Fifteen Categories of Marketing Data

Beneath the four quadrants of the Marketing Data Landscape sit at least fifteen core categories that define how modern marketing operates. The exhibit highlights the raw materials every CMO already recognizes, each carrying distinct opportunities and risks. While the mix varies by industry and company, these fifteen categories represent the most common and consequential forms of marketing data in use today.

Yet most analysis still happens in fragments. Teams work within segments, silos, or operating lanes, each optimized for its own definition of performance. That separation hides the real value that lives in the connections between categories.

AI changes this equation. By recognizing patterns across data categories, it unlocks insight that manual analysis rarely identifies.

It can model outcomes across the AIM OS, follow signals as they move in real time, and learn which actions actually improve performance. When intelligence connects these data sources, marketing data stops being something teams analyze after the fact and starts becoming something the business can act on while decisions still matter.

This is why the business goal isn't organizing data but making it usable. A list of data types tells you what you have. AIM OS shows how data moves and drives action. When these fifteen data sources are connected across the four zones, data stops sitting in place and starts flowing. Customer behavior triggers responses in real time, budgets adjust based on evidence, and trust is built into how decisions are made. For CEOs, this view shows where marketing investment creates leverage rather than overhead. For CMOs, it defines which data fuels decisions, protects the brand, and deserves the next round of improvement.

No dataset creates advantage on its own. Each performance zone depends on a different combination of data, and performance only compounds when those combinations are engineered to work together. Adaptive Decisioning relies on clean, unified signals to choose the right action. Autonomous Activation depends on timely context to execute at speed. Experience Loop Intelligence improves when feedback is captured and applied in real time. Intelligence Governance ensures that speed remains safe, trusted, and explainable. The exhibit shows the practical benefit of linking data directly to what each capability must do.

Exhibit 21: **How the Marketing Data Categories Map to AIM OS**

Capability Zones	How Data Is Used	Primary Data Inputs	Leadership Takeaway
Zone 1: **Adaptive Decisioning**	Data fuels real time decisions, targeting, and prediction.	Customer identity and profiles; Web and behavioral signals; Lead and account data; Third party and intent data; Purchase and transaction history; Revenue and pipeline data; Journey and attribution data	Better decisions come from clean, connected signals. Fragmented or low quality data makes AI fast, but inaccurate.
Zone 2: **Autonomous Activation**	Data drives automated execution, spend allocation, and triggered actions without manual intervention.	Campaign and engagement data; Advertising and media spend; Search and SEO data; Social and community data; Content performance	Automation magnifies whatever it touches. Poor data multiplies waste. Engineered data multiplies return.
Zone 3: **Experience Loop Intelligence**	Data links customer action to system response, so experiences improve continuously.	Campaign and engagement data; App and product usage; Content performance; Customer experience and service data; Purchase and transaction history; Journey and attribution data	Learning depends on speed. When signals move slowly or stay siloed, improvement stalls.
Zone 4: **Intelligence Governance**	Data sets trust boundaries and ensures customer confidence and compliance.	Consent, privacy, and governance data; Revenue and pipeline data; Journey and attribution data	Governance isn't a barrier. It's what makes speed safe, trusted, and sustainable.

Taken together, the AIM OS performance capability zones reveal both the breadth and the complexity of modern marketing data. Value appears only when data is designed to work together as a system. When it does, opportunity signals sharpen. Targeting improves. Behavior change becomes visible. And experiences feel personal rather than generic.

In teams we've worked with, data performance is rarely just a technical problem. It can also be a management issue. The challenge is hygiene, governance, trust, and alignment. That's why the next sections reorient the focus from datasets to discipline. This is the leadership work that turns data into sustained advantage.

Where Data Management Meets Leadership

The AIM OS capability zones reveal the full scope and complexity of modern marketing data. Performance comes from connection rather than collection. Technology creates potential, but leadership turns it into momentum. This section focuses on the leadership actions that make marketing data perform.

The eight imperatives that follow show how modern CMOs manage performance in motion. They explain how leaders turn information into movement, converting data assets into operating advantage. Each reflects a shift from owning data to orchestrating how it works across the AIM OS.

Let's take a closer look at the eight leadership imperatives.

1. Strategic Use of Data

The first discipline of marketing data is deciding what it is for. Too many organizations still treat data as exhaust. It's a byproduct of campaigns, channels, and tools. It accumulates in dashboards and warehouses, impressive in volume but unclear in purpose. Leaders assume more data will eventually translate into insight. But it rarely does.

The strategic shift is to treat data as intent driven signal context. For CEOs, the questions are straightforward. What decisions will this data enable? What risks will it reduce? What growth will it accelerate? For CMOs, the work is to design data collection and integration around those answers. This is the move from hoarding information to using AI to deliberately curate what is most likely to improve performance inside the AIM OS.

From exhaust to asset, there are three categories that illustrate the difference:

- *Customer identity and profiles.* Left unmanaged, identity data fragments into duplicates and conflicts. Once unified, it becomes the foundation for personalization and account based marketing.

- *Behavioral web data.* In isolation, it is often noisy and distorted by bots or low quality traffic. Combined with purchase and engagement signals, it becomes a meaningful indicator of pipeline momentum.

- *Third party and intent data.* Frequently purchased by default and rarely aligned to campaigns or sales motion. Used with purpose, it reveals which accounts are actively in market before competitors see the shift.

Each of these data types begins as a cost storage, enrichment, and management. When engineered with purpose, they compound into advantage. A second essential shift is abandoning the pursuit of perfect data. Perfect data doesn't exist. What matters is fitness for purpose.

In predictive modeling, a week of fresh behavioral signals can outweigh a decade of historical records. In personalization, a customer's most recent purchase is often more valuable than their entire history. That is why CMOs are reframing the question from "How complete is our data?" to "Good enough for what?" The standard moves from statistical perfection to strategic relevance. Data must be clean enough for models to decide, current enough to act, and trusted enough to engage the customer.

Leadership View: For CMOs, strategic use of data means managing two conversations at once. The first is upward. Frame the discussion with the CEO in capital terms. Which data assets justify continued investment because they enable growth and reduce risk, and which have become overhead that adds complexity without return.

The second conversation is inward. Focus teams on unifying identity, filtering noise from behavioral signals, and activating underused sources such as service transcripts or app usage so data changes decisions and behavior.

For CEOs, the takeaway is clarity. Marketing data is an enterprise asset. The question is whether it is engineered to create advantage inside the AIM OS or merely accumulated as overhead. Treat data as a designed marketing structure and it returns operating leverage. Treat it as storage and it returns clutter.

2. Value Creation in Practice

We believe data earns its keep when it creates tangible value like efficiency, growth, and risk reduction. CMOs are under constant pressure to prove that data is driving measurable performance. The challenge is that value doesn't emerge automatically. Data creates impact only when it is deliberately engineered into use cases that compound advantage inside the AIM OS.

Efficiency is often the first dividend. Campaign and engagement data reveal where budget is being wasted. Advertising and media spend data separates channels that deliver real ROI from those that generate vanity reach. Programmatic waste that once hid in the system becomes visible when signals are audited, modeled, and acted on. Each efficiency gain returns time, attention, and dollars that can be reinvested in market moves. That is how performance begins to compound.

Growth follows when data enables personalization and acceleration. Purchase and transaction records, once trapped in silos, reveal cross sell and upsell opportunities when unified across channels. App and product usage data surfaces churn risk early and highlights paths to expansion. Enriched lead and account data, paired with intent signals, allows sales to prioritize in market accounts before competitors see the shift.

Risk reduction is the third component. Consent and governance data keep campaigns on the right side of regulation. Service transcripts and chat logs surface reputational risk before it escalates. Trust is less visible than growth, but it compounds just as powerfully. You know that once brand trust erodes, it can take years to rebuild. When it's protected, it becomes a durable equity asset.

Leadership View: For CMOs, frame the discussion with the CEO in business terms. Which data categories reduce waste, accelerate growth, or protect trust? The inward work follows from that clarity. Guide teams to engineer marketing data into outcomes by unifying purchase records, activating intent at the moment it matters, and embedding consent checks into daily operations.

For CEOs, the question is simple. Are data investments delivering measurable efficiency, growth, and trust, or are they still trapped as overhead that adds cost without return?

3. Data Driven Decision Making

The third discipline is decision making turning data into choices that the business can act on. For years, marketing decisions have lagged customer behavior and surfaced after the fact in static dashboards. By the time attribution reports arrived, spend had already been misallocated and opportunities had passed.

That lag is now disappearing. Inside the AIM OS, agents can guide choices in near real time, shifting marketing from reactive to adaptive.

- *Attribution and journey data.* Once delivered as static reports, it can now be used in motion so budgets adjust before waste compounds.

- *Campaign and engagement data.* Previously limited to opens and clicks, it now feeds models that redirect spend toward high performing content while campaigns are still live.

- *Purchase and transaction data.* Formerly reviewed quarterly, it can be fused with behavioral signals to adjust pricing or promotion before churn accelerates.

This is what separates raw reporting from engineered decisioning. Data is no longer a rearview mirror. It becomes a guidance system. In practice, cycle time collapses from weeks to hours, sometimes minutes. The faster the system can sense and act, the more advantage it creates.

This discipline is also about teamwork and balance. Algorithms can surface options and reallocate resources in seconds. CMOs and their teams set guardrails, interpret tradeoffs, and decide when management judgment should override automated action. Decisioning works best when teams and agents amplify one another. Marketing Engineering designs that collaboration through specific thresholds and escalation paths.

Leadership View: For CMOs, decision making should be framed in capital terms. Show how decisions prevent waste, redirect spend toward growth and protect retention. Internally, organize teams around telemetry and models rather than static reports. Shift the question from "What happened?" to "What are we doing now?"

For CEOs, decisioning is an operating advantage, rather than a technology cost. Organizations that can reallocate spend and adjust offers faster, while keeping manual oversight light and deliberate, will outpace competitors still bound by calendar driven cycles.

4. Data Hygiene (The Non-Negotiable Foundation)

Every promise of data intelligence, personalization, and growth rests on a single foundation: hygiene. Without it, the AI system collapses. No matter which event, roundtable, or breakout you attend, the message is the same. Marketers already know this instinctively. Duplicate profiles inflate campaign counts. Bot traffic distorts engagement rates. Consent flags expire and expose the company to regulatory risk. Each issue may appear minor on its own, but together they corrode performance from the inside out.

> *That is why hygiene is the first discipline of Marketing Engineering and the entry condition for the AIM OS.*

Hygiene is about making data fit for purpose. Capture it cleanly, label it clearly, move it quickly, and keep the right permissions attached. Fix issues at the source. Resolve identity as signals arrive, filter bots before they distort results, and validate consent at capture. When hygiene is handled upstream, every decision becomes faster and safer. It becomes part of the AIM OS rather than a downstream cleanup task.

This reminds leaders that perfection isn't the standard. What matters is whether the data is good enough for the decision at hand. In an AI driven personalization engine, credibility may matter most. In attribution modeling, completeness and context carry more weight. The standard shifts by use case, but the principle holds. Relevance matters more than purity.

The impact of hygiene shows up in how systems are designed. Address it upstream where data enters and value multiplies. When identity duplicates are resolved at the source, every campaign and sales handoff runs cleaner. When bot traffic is filtered before it enters engagement streams, ROI models sharpen and budget decisions improve. When consent is validated at collection, compliance risk never reaches model training.

Leadership View: For CMOs, hygiene is both a leadership conversation and an operating discipline. Upward, frame it as a risk and return decision. Unchecked errors distort forecasts, erode trust, and create regulatory exposure. Inward, make hygiene part of how data is created and consumed. Use automation to catch and correct issues at scale, and management oversight to interpret anomalies and preserve confidence in the signals. Most important, treat hygiene as an ongoing engineering practice.

For CEOs, know that hygiene isn't an IT expense. It is the non-negotiable foundation of modern marketing performance. Without it, the AIM OS misfires, regardless of how advanced the AI or how large the dataset. Fix hygiene first, and everything else becomes easier to scale.

5. Engineering Data Systems

If hygiene is the foundation, engineering is what makes data usable at scale. Marketing leaders have learned the hard way that clean data alone doesn't create value unless it can move. From collection to decision. From signal to action. Fragmentation remains one of the biggest obstacles of performance.

For years, data integration meant one off projects and fragile pipelines. Campaign systems were connected to CRMs. Web logs were pushed into warehouses. APIs were stitched together in ways that broke whenever a vendor released an update. Each effort solved a local problem, but rarely produced a system that could adapt as marketing evolved.

The Marketing Engineering discipline changes that approach. It treats data architecture as structure versus a series of unrelated projects. That means designing data flows that are reusable, scalable, and resilient. It also means deliberately connecting marketing data into a system that powers the AIM OS.

A few design principles make the difference clear:

- *Identity resolution.* Customer profiles are continuously connected to behaviors and transactions in near real time versus being reconciled in monthly batches.

- *Consent and governance.* Privacy and compliance checks are built into data flows, so enforcement is automatic rather than retrofitted.

- *Activation systems.* Campaign orchestration, personalization engines, and sales and service handoffs draw from a shared data foundation instead of siloed databases.

Active collaboration and balance between teams and machine learning are essential here. Automation can move, clean, and enrich data at speeds no team can manually match. Marketing leaders and engineers decide which models matter, how systems connect, and how data flows align to business priorities.

Leadership View: For CMOs, engineering data systems requires two moves in parallel. Upward, frame data architecture for the CEO as structure. It reduces complexity and creates capacity for growth, rather than becoming a series of disconnected IT projects. Inward, push teams to think in flows and fusions instead of storage and reports. The work is to engineer data into an operating system for marketing, instead of a warehouse of unused records.

For CEOs, this system design determines whether marketing runs as an intelligent, compounding capability or remains trapped in one off integrations that never build advantage.

6. Attribution and Measurement

Few areas of marketing data generate more debate than attribution. For decades, leaders chased the idea of a perfect model. A single clean view showing how every dollar spent translated into every dollar earned. That goal is sound in theory. Marketing is far more complex.

Buying journeys are nonlinear. Consumer behaviors span channels and devices. And decisions unfold over time frames that no single model can fully capture.

Attribution still remains essential, but the standard is shifting. It's less about perfection and more about guidance that is good enough to act on. In practice, that means moving from static reports to dynamic measurement that learns and adapts over time inside the AIM OS.

Key shifts show how attribution has evolved:

- *Last touch models.* Once standard, but they're too simplistic to guide modern investment decisions.

- *Multi touch models.* These add nuance, but often collapse under data gaps and endless methodology debates.

- *Algorithmic models.* Fueled by real time signals, they can dynamically adjust spend, but only when underlying data is clean and connected across systems.

This is where attribution connects to measurement as an operating system function. Measurement is more than proving ROI to the board. It's how the system learns and improves. When engagement signals feed back into decisioning, targeting sharpens while it still matters. When conversion data flows into activation, spend shifts toward what is working and away from what isn't. When service and retention data enter the model, marketing aligns to lifetime value instead of short term wins. Used this way, measurement turns evidence into action and makes performance improvement repeatable.

Speed with judgment is the goal. AI models can process signals faster than any analyst. But it takes teams to interpret judgement, set guardrails, challenge assumptions, and monitor anomalies. Marketing Engineers ensure attribution frameworks align with strategy rather than simply amplifying noise.

Leadership View: For CMOs, engineered measurement should be framed around business impact. Show how it reduces waste and ties spend directly to outcomes the business cares about. Internally, organize teams around continuous feedback. Ensure every signal used for attribution also improves decisioning and activation, so campaigns sharpen with each cycle.

For CEOs, measurement is ultimately about control. When attribution is engineered well, it becomes an operating advantage inside the AIM OS.

7. Governance and Trust

If hygiene makes data usable, governance ensures it is used responsibly. In an AI driven operating model, where decisions move at unprecedented speed and scale, governance is paramount. It's the safeguard that keeps performance from becoming a liability.

Trust is the currency. And customers expect brands to use their data with discretion. Regulators expect companies to meet evolving standards for consent, privacy, and security. Boards expect leaders to identify and contain data risk before it damages reputation or enterprise value. Without governance, the same intelligence that fuels growth can erode trust just as quickly.

Governance goes beyond checklists. It embeds rules directly into the system. Consent is captured at the point of collection. This makes access to sensitive data role based and auditable. Retention is enforced by design, rather than manual cleanups. Leading teams are moving from policies on paper to governance as code, where rules are embedded in the data and compliance happens automatically inside the AIM OS.

A few focal points show where governance matters most:

- *Consent and privacy data.* Validate signals upstream so risk never enters the system.

- *Revenue and pipeline data.* Keep attribution and forecasting models transparent and auditable so outputs can be trusted.

- *AI models and agents.* Document training data, enforce guardrails, and monitor drift to prevent biased or opaque outcomes.

Team oversight here is indispensable. Automation can enforce rules at scale, but only our teams judgment can weigh gray areas and intervene when compliance tradeoffs collide with opportunity. Marketing Engineering manages this balance by defining actionable escalation paths, explainability standards, and ownership.

Leadership View: For CMOs, governance is both a boardroom topic and an operating discipline. Upward, frame it in terms of enterprise risk and reputation. Strong governance protects customers, the brand, and the balance sheet. Inward, embed governance into everyday workflows so compliance becomes part of the operating system. Work with legal, IT, and operations to place rules directly in the data layer.

For CEOs, governance isn't a brake on performance. It is what makes performance sustainable. The AIM OS compounds value when the system is trusted by customers, regulators, and the board.

8. Agent Ecosystems

The final discipline is the newest and, for many leaders, the least familiar. As generative models evolve into autonomous agents, marketing data moves beyond dashboards and analysts. It's consumed directly by agents that can act on their own.

This changes the role of data once again. Data becomes the operating fabric agents work within. A prospect profile, a consent flag, or a purchase signal becomes the trigger an agent uses to personalize outreach, launch programs, and adapt offers in real time.

In practical terms, the pressure shifts back to engineering once again. Agents require signals that are clean, current, and connected because they act without waiting for team review. They also need context. Metadata must tell agents what a signal is, but also what it means and what actions are permitted.

Early use cases make the shift tangible:

- *Intelligent routing agents.* They can qualify leads and hand them to sales in seconds, but only when identity and intent signals are accurate.

- *Campaign optimization agents.* They test and adjust creative dynamically, but only when engagement and spend data flow in real time.

- *Service agents.* They resolve issues instantly, but only when experience history and consent records are reliable.

Once again, the balance between teams and agents is essential. Agents act autonomously. CMOs define the guardrails. This includes which decisions are safe to automate, which require escalation, and how accountability is tracked when machines operate at scale. Marketing Engineers design this control layer with allow lists, confidence thresholds, audit trails, and rollback paths.

Leadership View: For CMOs, agent orchestration should be framed with the CEO as a question of strategic readiness. Organizations that engineer data for agents move faster, learn faster, and compete more effectively than those constrained by manual cycles. The question is whether the data is engineered to support autonomous action or whether the organization is exposed to errors made at machine speed. Ensure teams design, test, and govern these systems, training agents on trusted signals and triggering management intervention when confidence dips.

From Data to Engineered Marketing Advantage

The eight disciplines of marketing data, from strategic intent and hygiene to governance and agent ecosystems, may appear distinct. In reality, they function as one integrated operating model. Inside the AI Marketing Operating System, each capability zone depends on specific data, engineered to flow together with purpose. Design creates coherence, and coherence is what allows performance to compound.

When data is structured to move, governed to be trusted, and engineered to inform action, signals travel faster, decisions sharpen, and execution becomes more precise. Trust strengthens with every cycle because speed no longer comes at the expense of control. Marketing data stops being managed and starts being directed.

This is the shift from data accumulation to AIM OS advantage. From data as inventory to data as foundational infrastructure. When engineered with intent, marketing data becomes a leadership asset, powering adaptive decisions, scalable activation, continuous learning, and responsible growth. That is what separates modern marketing performance from activity. And it is what turns data into lasting competitive advantage.

Summary

If the data disciplines in this chapter feel extensive, that's expected. CMOs inherit years of layered workflows, vendors, and habits, none of which were designed to work as a single system. The goal isn't to master everything at once, but to run data as an engineered discipline inside the AI Marketing Operating System so each part strengthens the next. When data is designed with intent rather than accumulated by default, strategy becomes clearer, execution accelerates, and risk diminishes.

For the CMO, the mandate is to make intent visible. To show which data enables decisions, which reduces waste, and which protects trust.

Remember, perfection isn't the standard. Fitness for purpose is. Data must be clean enough to decide, current enough to act, and trusted enough to engage the customer. For CEOs, the frame shifts to capital. Marketing data moves from overhead to structure, and the proof shows up in falling cycle times, shrinking waste, and decisions made while opportunity still exists.

As AI advances, performance rises or falls on the quality and discipline of data. That demands clear ownership, the right skills, and sustained executive attention anchored in how marketing actually operates and how customers behave. It can't be handed off to IT or treated as a project with an end date. This is a permanent operating discipline. Because data must now be continuously refined as markets shift, behavior changes, and AI takes on more responsibility.

This work is achievable. It requires disciplined focus and steady progress. Marketing Engineering is the mechanism that makes this practical. It applies engineering rigor to how data, systems, and decisions operate, turning continuous improvement into habit rather than initiative. Done well, small changes compound. And the AIM OS strengthens over time.

Leadership Review

1. Are we treating marketing data as an enterprise asset or a scattered byproduct?

2. Do we have a coherent design for how data flows across the organization?

3. Do governance practices prevent errors upstream or absorb costs downstream?

4. What evidence shows that data is interoperable rather than locked in silos?

5. How prepared are we for autonomous AI agents acting on marketing data?

There is a practical limit to what human effort can extract from modern marketing data. No amount of manual analysis, reporting, or heroic work can uncover the value hidden across dozens of fragmented data sources at speed or at scale. That limit is structural and it's why marketing performance now depends on systems. For CMOs, this should be grounding. The complexity is real. The challenge isn't lack of insight or discipline, but years of growth, tools, and workflows layered without a unifying architecture. This is where AI becomes essential. AI unlocks the foresight embedded across this data landscape, revealing patterns, dependencies, and opportunities no team could surface manually. Seeing the full picture is the first step toward engineering that advantage deliberately.

Exhibit 22: Marketing Data Landscape

Category	Why It Matters
Customer Identity & Profile Data	Fragmented identity creates duplicates and conflicting records. Unified identity enables personalization, ABM, and clean handoffs across sales and service.
Behavioral Web Data	Bot traffic and noise distort engagement. Clean behavioral streams surface real intent and improve opportunity detection.
App & Product Usage Data	Declining usage often signals churn before renewal risk is visible. When monitored, it enables early intervention and expansion.
Campaign & Engagement Data	Privacy changes degrade opens and clicks. Shifting focus to conversion and action improves signal quality.
Lead & Account Data	Over scoring inflates pipeline and misleads forecasts. Tuned scoring sharpens account focus and improves conversion.

Purchase & Transaction Data	Siloed commerce and retail data hides lifetime value. Unified views reveal cross sell and upsell opportunities.
Attribution & Journey Data	Last click rules obscure long buying journeys. Multi-touch visibility improves investment decisions.
Search & SEO Data	AI summaries reduce clicks. Influence shifts to shaping AI driven discovery and intent.
Social & Community Data	Vanity metrics mislead performance. Sentiment trends predict churn, advocacy, and brand risk.
Advertising & Media Spend Data	Programmatic waste drains ROI. AI driven targeting improves efficiency and return.
Content Performance Data	Overproduction without consumption wastes resources. Usage insights guide high-impact content.
Customer Experience & Service Data	Disconnected systems hide dissatisfaction. Integrated signals flag churn and recovery opportunities early.
Third-Party & Intent Data	Often underused by default. When aligned, it identifies accounts already in market.
Revenue & Pipeline Data	Poor linkage between marketing and revenue clouds ROI. Alignment restores credibility and forecast confidence.
Consent, Privacy & Governance Data	Stale consent erodes trust and creates risk. Transparent governance strengthens confidence and compliance.

12

Engineering Marketing Efficiency for Velocity

*How operational efficiency creates
the context for AI performance*

Space exploration may seem like an unlikely place to begin a chapter on marketing, but the shift that made rockets cheaper, faster, and smarter mirrors the challenge marketing leaders face today. The goal is to stop treating efficiency as cost cutting and start using it as thrust for modern marketing.

Over the past decade, space exploration has shifted from rare, expensive launches to frequent, learnable cycles. What once required government budgets and decade-long timelines is now a competitive arena where private companies cut costs, shorten cycles, and scale ambition.

Consider the traditional space exploration model. According to NASA, the Space Shuttle program averaged roughly $1.6 billion per launch when development, operations, and infrastructure were included. Launches were infrequent. Hardware was discarded after each mission. And improvements waited for the next launch window, often months or years away. Experimentation was very expensive, which made learning and innovation slow.

Then a new wave of companies applied a shared set of principles to make launches cheaper, faster, and repeatable. Organizations like SpaceX, Blue Origin, Rocket Lab, Firefly Aerospace, and Intuitive Machines reengineered the economics of exploration. Reusable systems, tighter vertical integration, and streamlined supply chains collapsed launch cycles from years to weeks. Each flight became both a mission and a learning event.

Across the growing sector, the lesson is consistent. When operational efficiency lowers the cost per launch, experimentation accelerates. More launches produce faster learning, tighter feedback, and compounding innovation. Efficiency stops being a constraint and becomes the engine of progress.

The impact on the U.S. space program has been transformative. Mission volume and diversity surged, from satellite constellations to lunar landers. Countries without this operating model struggle to match U.S. launch frequency, turnaround speed, and cost efficiency. That is the standard marketing leaders should pursue. Modern marketing wins by moving faster, learning faster, and competing through a system designed to repeat success.

The lesson goes far beyond merely cost reduction. The entire innovation cycle accelerated. Innovations like reusability, vertical integration, and streamlined operations collectively reimagined turnaround times. These faster cycles led to more missions, more learning, and continuous improvement. Operational excellence became the catalyst for innovation.

Efficiency in this context isn't about doing less.
It creates capacity for doing more.

This mindset translates directly to marketing. Efficiency creates capacity for more launches, broader coverage, and higher revenue. It frees resources to pursue market opportunities, test ideas, and scale the ones that work. That is why this case study matters. Efficiency isn't a constraint. It's a multiplier of performance.

This same dynamic is unfolding in modern marketing. The AI Marketing Operating System can dramatically increase how quickly and precisely teams respond to customer signals, but only when the foundations are ready. Without clean data, tighter workflows, and a tuned martech stack, AI won't solve your inefficiency. It will accelerate it. Just as lower launch costs reshaped aerospace, optimizing the systems and structures you already own will reshape marketing. Efficiency creates the conditions for intelligence to perform, enabling faster cycles, bolder experimentation, and more ambitious growth. Marketing organizations that strengthen their foundations now will outpace those that wait.

For too long, marketing has treated efficiency as austerity. Budget cuts, hiring freezes, and mandates for marginal savings created constraints. In the AI era, efficiency means the opposite. It's the platform that multiplies marketing's ability to act, adapt, and innovate at speed. Efficiency is no longer cost control. It's competitive fuel. Leaders who engineer efficiency create the launchpad for faster learning, greater scale, and sustained growth.

We Have Competitive Separation

When a system runs faster and leaner, teams can attempt more without exhausting resources. This is where efficiency becomes a platform for competitive advantage. In the global space economy, it enabled mission scale at a pace once considered impossible. The table below shows the competitive separation the U.S. has achieved in launch volume in recent years. The leadership question for CMOs is straightforward: *What would it take to create this level of competitive separation for your brand's market position?*

Exhibit 23: **Orbital Launch by Country (2016-2025)**
Source: Our World In Data, CSIS Aerospace, Pay Load Space

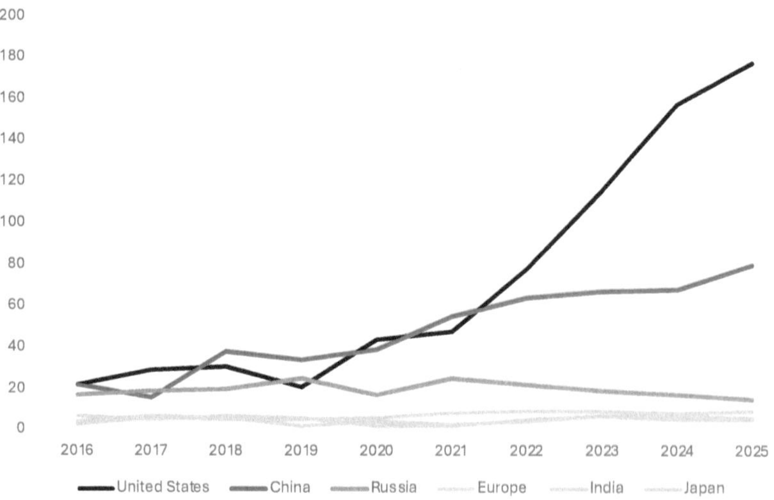

Operational excellence is a catalyst for competitive performance. Stacking new AI capabilities on top of inefficient, siloed systems isn't a sustainable strategy for modern marketing. The same shift that transformed the space economy is now reshaping marketing. The AIM OS plays the role of reusable rockets. When engineered properly, it lowers the cost of action and improves with every cycle.

The goal isn't cost cutting. It's moving from complexity to a streamlined foundation that enables faster execution and stronger competitive performance. In practice, this means modernizing systems, eliminating redundancy, and tightening workflows so every dollar saved becomes a dollar invested in market moves.

Most marketing technology structures aren't ready for the AI era. Data is fragmented and incomplete, limiting its value for AI driven decisions. Martech stacks are bloated with overlapping tools and weak integration. Workflows are loosely documented, inconsistently followed, or built for manual execution. AI is treated as a side project, focused on content creation rather than decisioning and orchestration. Over time, these elements have been stitched together by necessity, instead of by design. Fixing this is more than a tactical cleanup. It's foundational engineering and it is strategic work.

Marketing Engineering exists to untangle that patchwork and restructure a cohesive, high-performance system.

It's like trying to run a reusable rocket program with mismatched parts and outdated launch systems. You cannot scale innovative missions until the fundamentals are fixed.

And the clock is ticking. AI adoption is accelerating, and the gap between AI ready systems and inefficient ones is widening fast. Every month, competitors learn faster and unlock advantage. If your foundation isn't fit for modernization, AI won't close the gap. It will magnify inefficiency across your teams.

What's at stake if these issues remain unresolved is market tempo. Competitors with cleaner, faster systems will run multiple campaigns in the time it takes you to run one. They'll act on signals while they're still relevant, rather than weeks later after opportunity has passed. They'll test more ideas, refine more messages, and scale winners before you even see them coming.

Over time, more efficient competitors will uncover more segments, refine product offers faster, and own more of the market conversation. By the time you clear backlogs or fix the latest bottleneck, the audience has already moved on. The AI era rewards speed, and lost market cycles are rarely recoverable.

In aerospace, cheaper and more reliable launches unlocked a new pace of innovation. In marketing, a clean foundation does the same. Efficiency is the foundation that creates faster cycles, more experimentation, and bolder moves. Without it, AI remains a novelty. It creates impressive pilots but disappoints at scale. With efficiency, marketing gains a launch system that fires repeatedly, learns quickly, and adapts faster than competitors can respond.

I recently met with a CMO who was energized about the AI future. He called it their "moonshot moment," and the analogy fit. Predictive analytics, generative content, and advanced segmentation were already underway. Budgets were committed. Vendors were engaged. And momentum felt real.

But he also described cracks in the foundation. The customer data platform (CDP) still carried unresolved CRM issues. Data from multiple martech tools had to be assembled by hand, with no integration plan in scope. Campaign automation workflows were discussed endlessly across teams, yet rarely moved forward. These weren't new problems. They're the accumulated result of years of layered systems and shifting priorities.

When I asked how these issues would be fixed before the AI pilots rolled out, his answer surprised me. "AI will help us sort that out." That way of thinking is a trap. AI doesn't clean the house for you. If data is messy, the stack is uncalibrated, and workflows are slow, AI simply produces bad outputs faster. You end up with a high-speed version of today's mess. The real moonshot isn't launching AI pilots. It is readying the launch system so AI can fly farther and faster.

Without foundational readiness, AI becomes an
expensive experiment, versus a performance breakthrough.

Most marketing organizations don't start with a clean slate. They inherit a patchwork of systems bought for past priorities and processes built for another era. This creates an operational barrier. And overcoming it requires Marketing Engineers who understand both modern performance demands and the legacy layers being transformed. Skip this work and the AIM OS will not build speed. Instead, it will compound instability.

Three Essential Foundational Fixes

For the AI Marketing Operating System to perform at scale, its foundation must be sound. That requires addressing three sources of operational drag before expanding AI use. These aren't back-office cleanups.

They're the structural levers that determine whether AI remains a series of pilots or becomes a true performance multiplier. Together, they form the core of the AIM OS introduced in Chapter 5, where data, martech, and AI driven workflows converge to drive modern performance. If any element is weak, the system cannot deliver its full potential.

1. Data: The fuel supply

Data is the lifeblood of AI, yet in most marketing organizations it doesn't flow cleanly. Duplicated records, fragmented customer histories, broken attribution, unreliable sources, and inconsistent field definitions undermine AI's ability to make accurate, timely decisions. Before models can be trusted, the fuel must be unified and refined. That means mapping sources, resolving duplicates, filling critical gaps, and enforcing governance that keeps data clean as it evolves. This is not a one-time cleanup. Data engineering is a continuous discipline, and its importance only grows as AI takes on more responsibility.

2. Martech Stack: The launch vehicle

Your martech stack is the vehicle that carries AI into action. If it's weighed down by redundancies, weak integration, or unused capabilities, it will never reach orbit. Many stacks today are bloated with overlapping tools, idle features, and point solutions that fail to connect to a coherent operating model. The fix is almost never martech expansion but streamlining for precision and reliability. Audit every platform, retire what doesn't serve a defined role, and fully integrate what remains into the AIM OS and business objectives. Every platform must earn its place. If it doesn't contribute to the system, it doesn't belong in the stack.

Martech requires ongoing governance and calibration. Without active management, features go unused, capabilities drift, and investments depreciate. What should accelerate performance quickly becomes a liability. Marketing leaders treat the stack as a dynamic system, continuously tuned to protect investment and sustain speed.

3. Workflow: Mission operations

Even with clean data and a tuned stack, performance breaks down if workflows lag. This is the signal-to-action layer. The clock starts when a market signal appears and stops when a response reaches the customer. In too many organizations, approvals, handoffs, and manual steps stretch that window from hours into weeks.

Delays often come from good intentions. Teams chase perfection, wait for executive feedback, or debate creative nuance while momentum drains away. In the AI era, that delay is fatal because insights fade by the hour. I have created these bottlenecks myself by letting message platforms and creative cycles run too long. The lesson is simple. Speed is now part of quality.

The fix is to remove friction. Automate wherever possible, push decisions closer to the front line, and eliminate steps that slow AI driven action. These three fixes form the foundation of the AIM OS, where data, martech, and workflows converge. If any layer is weak, the operating system misfires and performance suffers.

Modernizing the foundation requires
the discipline of Marketing Engineering.

Marketing Engineers understand modern system architecture, AI integration, and how to design for performance from the ground up. Don't delegate this work to generalists or leave it to IT alone. If these foundations falter, the entire system is at risk. Leaders who place accountable Marketing Engineering and ownership on strong foundations and manage them as a priority, gain speed and reliability when it matters most.

This work must be elevated and prioritized. AI adoption is accelerating faster than any prior transformation, and every delay compounds the performance gap. Competitors who move now will have tuned systems feeding real time models long before slower organizations clear their backlog. One of the best indicators of AI performance maturity we see isn't the range of pilots or programs, but the assessed strength of these three foundational levers.

From Launch Cost to Launch Velocity

In traditional marketing, efficiency meant cost cutting and doing the same work with fewer resources. In the AI era, efficiency plays a different role. It becomes the platform for velocity. Every source of friction removed from data, stack, and workflows increases how fast the AIM OS can sense, decide, allocate, and act.

This is how efficiency multiplies performance. Lowering the cost per launch doesn't just save budget. It expands your capacity to reach the market more often, with greater precision, and with learning built into every round.

Operational efficiency is how teams
go faster without draining resources.

That's why, in AIM OS, efficiency becomes the foundation for acceleration. It replaces a few large, risky bets each quarter with many smaller, faster moves that compound momentum. It's the difference between chasing leaders and setting the pace.

When the foundations are in place, velocity becomes self-sustaining. Each cycle strengthens the system, just as frequent launches in space produce better hardware, faster turnarounds, and more ambitious missions. AIM OS works the same way. Every interaction delivers an outcome and feeds the next, with increasing precision.

As the space economy evolved, lower launch costs enabled more missions, more experimentation, and entirely new ambitions. In marketing, freed resources can be reinvested in next generation AI strategies, deeper personalization, and bolder moves that once felt too costly. Engineering efficiency becomes the platform that scales ambition.

The Competitive Advantage

The U.S. surged ahead in the modern space economy not just through better technology, but through systems that launched more often, adapted faster, and improved with every cycle. The same dynamic will separate winners from laggards in AI powered marketing. Speed, adaptability, and iteration now define advantage. Organizations that launch frequently, learn from each round, and feed insights back into the system set the pace for the market. Efficiency frees budget and bandwidth for more ambitious moves. Those who treat efficiency as a multiplier pull ahead. Those who treat it as an afterthought (or financial mandate) lose thrust while competitors are already airborne.

For CEOs, the implications are direct. Efficiency becomes a financial lever. A marketing organization designed for velocity delivers higher return per dollar, faster signal-to-action time, and lower marginal cost of growth. Throughput increases without proportional headcount. Learning compounds quarter after quarter. Instead of incremental lift from isolated campaigns, the enterprise gains structural performance that scales growth at lower cost.

The same principle appears in every high-performance field. In racing, teams rebuild cars between events to capture marginal gains that compound across a season. In manufacturing, leaders track line speed and defect rates continuously. In logistics, fleets are optimized in real time to eliminate waste before it accumulates. Winning systems combine velocity with reliability. That is the role of the AI Marketing Operating System and the discipline of Marketing Engineering. Together, they turn efficiency into advantage by making launches frequent and performance repeatable.

Readiness Check: Go/No-Go Sequence

This chapter has focused on three foundational fixes: data, martech, and workflows. Together, they form the base of the AIM OS. We've also been explicit that this work belongs to Marketing Engineering, versus marketing generalists or IT acting alone. This readiness check verifies whether your system is truly engineered for the shift to modern marketing.

Before launching AI pilots or adding new platforms, one principle matters most. And that is, AI amplifies what already exists, which is why fragmented CDPs and ineffective data workflows create failure points. If the system is fragmented, AI multiplies inefficiency. If it is tuned, AI multiplies performance.

This is why every organization needs a Go/No-Go sequence before liftoff. Rockets don't launch without passing system checks. Marketing should be no different. Use the ten tests in Exhibit 24 to determine whether you are flight ready or still carrying operational drag.

Exhibit 24: **System Go/No-Go—
Ten Launch Tests for Modern Marketing**
(Score each item from 1 to 5 to assess readiness)

1. *Data quality*: Is customer and performance data clean, unified, and actively managed for accuracy and timeliness?

2. *Data access*: Can teams reach and apply the right data at the moment of need without delay or manual workarounds?

3. *Signal flow*: Can a customer signal reliably trigger an in-market action in hours or days, and not weeks?

4. *Stack integration*: Is the martech stack integrated end to end, with redundancies eliminated and a specific connection to business goals?

5. *Stack management*: Is the stack actively managed and calibrated, with features enabled, teams trained, and unused components retired?

6. *Workflow speed*: Do initiatives move from concept to execution without recurring approvals or bottlenecks?

7. *Workflow clarity*: Are workflows documented, visible, and repeatable rather than dependent on individuals?

8. *Talent fit*: Do we have Marketing Engineering talent accountable for managing and evolving marketing technology and data foundations?

9. *Leadership commitment*: Is there executive leadership commitment to foundational cleanup before scaling AI initiatives?

10. *Efficiency discipline*: Are efficiency gains measured and reinvested to fuel growth rather than treated as one-time budget cuts?

How to Evaluate Your Readiness

For each question, score your organization from 1 (low readiness, significant gaps) to 5 (fully ready, no major gaps). Patterns of low scores reveal systemic weaknesses and should become your highest priorities for action. A total score of 40 or higher indicates structural readiness to accelerate into the AI era. A score below 30 signals urgent foundational work before AI compounds existing problems. Treat this as a team exercise, instead of a solo judgment. Multiple perspectives surface blind spots that no single viewpoint can see.

The three cleanup zones aren't housekeeping tasks. They are the domain of Marketing Engineering and the base layers of the AIM OS. Only when these foundations are strong can the four capability zones (Adaptive Decisioning, Autonomous Activation, Experience Loop Intelligence, and Intelligence Governance) operate as a connected, high-performance system. Weak foundations slow everything above them. Every efficiency gained here unlocks deeper AIM OS capability and faster compounding performance.

Readiness is more than just a technical issue, it's also cultural. Marketing Engineering goes beyond managing projects or delivering campaigns. Advantage comes from a culture of system improvement, where every initiative leaves the system stronger and smarter. Leaders must reward iteration and learning as much as visible wins. The question shifts from "How did this program perform?" to "What did it teach us, and how will it improve the next?" This shift is what prevents organizations from sliding back into outdated legacy habits.

The modern space economy rewards systems that launch more often, more reliably, and more intelligently. The AI era will do the same for marketing. The speed of your AIM OS won't be limited by ideas. It will be limited by whether the foundations can carry them forward.

Winners won't be the teams that experiment with AI. They'll be the ones that embed efficiency into the system from the start. These organizations cycle faster, capture more market moments, and scale what works while slower rivals are still fixing the last campaign.

Clean your marketing launch pad now or fall behind later.

In a market built for speed, every missed cycle is gone for good. We like to advise that the AI era is not about being first to adopt. It is about being first to adapt. With strong foundations, AI accelerates performance. Without them, it distracts. And in this market, distraction is another word for defeat.

When the next budget cycle arrives, will you show the output of a tuned, high-velocity system, or explain why AI pilots never scaled? The difference is more than vision. It's whether the system was engineered in time.

Summary

Efficiency has never been the endgame in marketing. In the AI era, it is the unlock. Not because efficiency cuts costs, but because it compounds motion. The lesson from the modern space economy is very relevant. Lower the cost per attempt and you get more missions, more learning, and faster progress. The same dynamic now governs marketing. Every inefficiency removed creates capacity for more attempts, sharper adjustment, and quicker wins. Efficiency isn't doing more with less. It is building structural capacity for acceleration.

In past eras, marketing could survive inefficiency because cycles were long, and expectations were forgiving. That time is over. Drag anywhere in the system, whether messy data, undermanaged stacks, or stalled workflows, doesn't stay contained. It spreads through everything AI touches and turns small problems into expensive ones. Fix the roots and momentum starts to compound.

This is why the three foundational fixes matter. Clean data enables trustworthy decisions. A tuned stack executes instructions reliably at scale. Fast workflows convert signals into action before the moment passes. Together, they turn the AI Marketing Operating System from promise into performance. When the foundation is sound, velocity becomes repeatable and results compound by design.

Marketing Engineering is the discipline that makes this real. It connects data, martech, and operations into one operating rhythm. It converts efficiency gains into capacity for more attempts, more learning, and more wins. That is how efficiency becomes strategy. It's not housekeeping or optimization. It's the basis of competitive advantage in modern marketing.

If you want AI to accelerate growth, create the modern marketing foundation to go fast. Clean the launch pad. Calibrate the system. Then let efficiency create lift and expand possibilities.

Leadership Review

1. Are we treating efficiency as cost reduction, or as capacity creation?

2. Which reusable booster moves will fund modern marketing performance?

3. Where does system drag exist that AI would only accelerate?

4. Do we measure efficiency by savings alone, or by the growth capacity it unlocks?

5. Do we pass the System Go / No Go Ten Launch Test?

PART IV

Marketing Leadership in the AI Era

13

Breaking Momentum Barriers

*Avoid compromises that
pull marketing teams backwards*

We once ran a workshop with a marketing leadership team to surface the risks that could slow their shift toward an AI driven model. In less than an hour, the group identified twenty-seven potential issues. We organized and grouped them into eight core challenges the team felt ready to tackle.

During the break, the CMO joked, "My team could probably find twenty-seven issues with ordering lunch." He meant it lightly, but it revealed something important. His team cared enough to probe every angle. His team wasn't being cynical. They were engaged enough to examine the problem from every angle. I told him it showed the team understood the landscape and could see where actions mattered. Real progress happens when issues are visible enough to be prioritized and addressed.

He paused, nodded and agreed, "Yes, blind spots kill momentum." That exchange captured the real challenge of moving to an AI driven model. For CMOs, risk isn't ambition or effort. It's failing to see the barriers before progress stalls.

Every paradigm shift reaches a decision point. Leaders either commit to a new operating model or keep trying to extract the last bit of value from the old one. For marketing leaders, that moment has arrived. The shift to the AI Marketing Operating System isn't a tune up. It's a structural overhaul of how decisions are made, how experiences are delivered, and how growth is sustained. It's about rebuilding marketing for an entirely new performance reality.

Every quarter spent optimizing an outdated structure is a quarter competitors spend training their AI systems and widening the gap. Once that lead is established, it can't be closed by pushing harder on yesterday's methods. Recognizing the need to change isn't the same as executing it. Marketing history is full of teams that launched ambitious agendas only to drift back into old habits. They rarely fail for lack of ambition. They fail because predictable compromises slowly erode conviction and momentum. This can be avoided.

This chapter identifies 15 barriers that we see stall progress and keep marketing anchored to legacy performance models.

Stories From the Field

Progress in marketing rarely stalls because of one major decision. It slips away through smaller moments that feel harmless at the time. A delayed approval. A budget trimmed. A process left untouched because it feels familiar. These are the quiet forces that slow momentum long before anyone calls it a failure. The stories that follow come from leaders who have lived this reality and show how easily progress can stall, even when intentions are strong.

The zero-based budget curveball. A CMO had full sponsorship for the shift to AI. Then the next budget cycle hit. A zero-based review stripped committed funding and forced every expense to be rejustified. The team was pushed into a false choice between running today's operations and building tomorrow's system. Momentum stalled before the new model had a chance to prove itself.

The pull of business as usual. Another CMO began with a plan for adoption, then gradually bent it to fit existing processes. Each compromise made internal alignment easier, but it diluted the redesign. By the time the organization was ready to scale, competitors were already showcasing their AI innovations.

The lost-in-translation delay. A leader admitted their AI initiative stalled in IT's queue. Marketing and IT spoke different operational languages, so priorities drifted and timelines stretched. The problem wasn't technology. It was surrendering operational fluency to the IT function.

The agency blind spot. A CMO asked an agency to redesign operations. The agency knew the deliverables, but lacked experience working inside a marketing organization. Progress slowed, and the client eventually paused to find partners with real operating experience.

The "this will pass" mindset. At an industry event, many hands went up when asked who was skeptical about AI in marketing. Several saw it as a threat rather than an opportunity. One CMO said, "If my team felt like that, we'd never adopt it." The problem wasn't just a lack of training. It was the failure to make AI personally relevant to the team's recognition and advancement.

These stories show that progress rarely collapses because of one sweeping leadership decision. More often, it erodes through a series of small compromises that feel harmless at the time. The shift to AIM OS is a pivotal change in the operating model, and it demands more awareness and discipline than most leaders expect.

This is the moment when marketing leaders must stay alert to roadblocks and read the landscape as it shifts. Not every adjustment is a setback, but every decision carries weight. Collaboration requires negotiation, yet each small concession compounds. One skipped checkpoint. One deferred investment. One "we'll get to it next quarter." Over time, these choices slowly move the function off course.

The 15 Momentum Barriers

The journey to modern marketing systems rarely stalls because leaders lack ambition. It stalls when momentum meets recurring issues. These compromises live inside the normal day-to-day operating cadence. A well-intentioned budget cut. A quick platform change. A shortcut taken before data is ready. Each feels reasonable in isolation.

Some barriers are structural, embedded in operating models that feel safe but slow progress. Others are technical, like data that works for reporting but breaks when AI depends on it. Others are cultural, habits and incentives that pull teams back toward the comfort of familiar ways of working. Avoiding these barriers requires more than awareness, like reshaping the culture and conditions that allow progress to compound rather than stall.

Most operating barriers are easy to avoid once you recognize them.

The fifteen barriers that follow come directly from what leaders encounter as they move from traditional marketing to the AIM OS. They are recurring patterns that consistently drain speed and confidence if left unchecked. Each issue includes indicators and early signals that momentum may be slipping. For clarity, the barriers are grouped into three categories: Strategy and Operating Barriers, Technology Adoption Barriers, and Talent and Partner Barriers. Use them to help your team see the field clearly, manage risk deliberately, and keep progress moving forward. Let's dive in:

Category One: Strategy and Operating Barriers

These compromises appear when the operating model lacks a deliberate end state, pace, or metrics. Without those anchors, organizations default to optimizing legacy activity instead of building AIM OS performance.

1) Clinging to Traditional Marketing Approaches

When budgets tighten or risk tolerance drops, leaders often return to what feels safe. Familiar methods get refined even though they were built for slower cycles and linear execution. Each quarter spent optimizing the old playbook is a quarter competitors spend building systems that scale and learn faster with every interaction inside AIM OS. The risk is subtle. This issue can look like discipline. What feels wise in the moment becomes expensive over time because the system never develops the adaptability modern growth requires.

Indicators: Strategy updates emphasize efficiency gains over future-ready capability. Budgets protect incremental upgrades instead of AIM OS alignment. Core programs still depend on manual workflows and calendar-based launches.

Solution: Replace one calendar-driven campaign with a signal-triggered program this quarter. Measure signal-to-action time and first-pass effectiveness. Protect a dedicated budget line for system work, tie it to executive KPIs, and hold the line.

2) Failing to Define the Vision Clearly

When leaders commit to a shift without a clearly defined vision, the organization defaults to incremental choices. Partners are selected on price. Roadmaps bend to convenience. Success gets measured on mechanics instead of market impact. The work fragments into disconnected projects rather than forming a unified system. Without a shared end state, teams optimize for the short term and performance never compounds. Competitors with a visible vision move faster, attract stronger talent, and pull away.

Indicators: Strategy plans set targets but never define the system end state. Budget debates focus on allocations rather than outcomes. Execution teams chase quick wins that don't connect to the operating model.

Solutions: Write a concise AIM OS end-state summary. Describe how decisions are made, how activation runs, how learning feeds back, and how governance keeps speed safe. Review it monthly at the executive table and before major spend decisions. Define three must-wins for the operating model and track effectiveness against them.

3) Underestimating the Speed of Change

Markets, customer expectations, and technology cycles are moving faster than most leadership teams plan for. Waiting for perfect conditions means starting behind and staying there. This kind of caution often feels responsible, but the delay is costly. By the time a safe plan is finalized, faster competitors have already adapted to new conditions and reset the performance bar.

Indicators: Competitors launch AI enabled initiatives while your team is still assessing options. Adoption roadmaps wait for a complete plan instead of advancing through iteration. Leadership discussions emphasize minimizing disruption rather than accelerating advantage.

Solution: Shift from planning to progressive readiness. Launch one controlled AIM OS flow within sixty days, focused on a narrow workstream where data, technology, and workflow can align quickly enough to learn. Replace the language of "pilots" with "learning sequences" so experimentation becomes normal operating behavior, rather than a special exception. In the AI era, waiting for certainty has become a new form of risk.

4) Defining Strategy by Outputs Instead of System Performance

Many strategies list outputs like engagement, pipeline, revenue, or ROI. Targets set ambition, but they don't define how the system must run to achieve them. Those results rely on traditional workflows, so the model never changes. An AI driven system should be judged by *how* it adapts, learns, and compounds advantage. Without that view, leaders chase numbers while competitors evolve their systems.

Indicators: Plans set goals but never specify how the system will learn and adapt. Leadership reviews treat AI as a tool to boost traditional metrics rather than a redesign of operations. Dashboards measure activity but ignore the system performance metrics.

Solution: Add a system KPI layer to the strategy and report it first. Track signal-to-action time, first-pass effectiveness, and automation coverage. Tie budget releases to improvements in these measures. Require every program plan to state what the system will learn and how that learning will improve the next decision.

5) Overreliance on Following

In traditional marketing, waiting to see what works for others could feel like a necessary hedge. In AIM OS, that logic breaks down. Early adopters don't just test new features. They build learning flows, train models, and refine operations in ways that compound quarter after quarter. Once a competitor's system is learning at scale, imitation can't close the gap. Each cycle makes them faster and more precise while followers remain static. By the time a fast follower deploys, the leader has already monetized years of proprietary signals and insights that can't be copied.

Indicators: Strategy decks cite analysts and vendors more than internal signals and outcomes. Leadership delays decisions while waiting for external proof. Teams pause for external validation instead of testing and learning in live flows.

Solution: Replace external proof with internal evidence. Set a ninety day goal to generate one internal case study that shows measurable lift from AIM OS in action. Choose a data driven program or an AI assisted process and run it end to end. Document what changed, what failed, and what improved. Share the case internally so teams can see progress and build confidence.

Category Two: Foundation and Speed Barriers

Predictable issues emerge when data, technology, or orchestration breaks down. When customer signals aren't clean, connected, and current, decisions fail to reach the market at speed and AI amplifies noise instead of performance.

6) *Assuming Your Data Is Ready*
Dashboards can create a false sense of progress. Data that's good enough for reporting is rarely good enough for AI. Intelligent systems require signals that are unified, governed, and enriched so learning can occur across interactions. Models trained on inconsistent, duplicate, or incomplete records produce noise, when intelligence is required.

Indicators: Planning cycles still rely on manual pulls across systems. Targeting and personalization fields are incomplete or inconsistent. Consent and provenance are unclear at the record level. Data freshness varies by source with no defined service levels.

Solution: Run a weekly identity and consent check on priority segments. Pause activation when identity or eligibility is unclear and fix issues at the source. Capture consent at the moment of collection and maintain an auditable trail. Set freshness targets for critical tables and set alerts for drift. Define a single golden record for identity and eligibility and enforce it through data contracts across platforms. Track three key metrics: identity match rate, consent validity rate, and field completeness.

7) *Layering Martech Without Reconciliation*
Adding tools without reconciling how they fit into AIM OS creates clutter instead of capability. Stacks expand, workflows overlap, and responsibilities blur, but the system doesn't move closer to intelligence at speed. Stack expansion gets mistaken for evolution. Without alignment to the AI operating vision, each new layer adds complexity, and the signal-to-execution flow fills with noise.

Indicators: Tools are added without a defined role in AIM OS. Stack growth outpaces the team's ability to configure, train, and calibrate. Workflows and data fail to connect cleanly across platforms. Feature adoption remains low while license counts rise. Signal-to-activation time stays flat or worsens despite increased tooling.

Solution: Build a capability map aligned to the AIM OS zones (Chapter 9). For every platform, define its role, owner, required inputs, outputs, and the handoff it supports. If a tool doesn't advance a named zone, reassess its place in the stack. Establish integration contracts and publish a quarterly martech health scorecard with four measures: utilization rate, feature adoption, redundancy flags, and decision-to-activation time. Tie renewal decisions to performance against these measures.

8) *Using Old vs New Metrics in Leadership Updates*
Too many leadership updates stay anchored to safe and familiar metrics. The update becomes a ritual instead of a steering mechanism. In AI driven marketing, leadership updates must function as live system health checks.

Indicators: Executive updates highlight campaign results but omit system progress or emerging risk. Wins are held back for reveals instead of shared as learning in motion. AIM OS metrics focus on outputs while operating and telemetry measures are absent.

Solution: Lead every executive update with a one-page AIM OS Health summary and make it the first agenda item. Show trend movement, include "because" for each change, and name an owner so action stays well defined. Add one monthly "learned and promoted" item where a test becomes the new standard. Replace vague risk language with defined thresholds and key dates. Keep this review standing so progress becomes an operating rhythm.

9) *Piloting Without a Path to Scale*
Pilots can prove concepts, but they often remain isolated. Early results are celebrated, and then momentum fades. Without specific

ownership, resourcing, and funding for rollout, performance never scales into mainstream operations.

Indicators: Budgets stop at testing with no allocation for deployment. Pilot teams disband with no named owner for adoption. Integration plans for data, workflow, governance, or support are missing. Results cannot be reproduced outside the pilot team or channel.

Solution: Define a scale plan before the pilot begins. Fund a small AIM OS venture budget that unlocks only if system KPIs are met. Keep it simple with one page covering owner, integration scope, timeline, and the system measures that must improve at scale. Require a decision within thirty days of the pilot readout. Keep the pilot team intact through first deployment, then assign a permanent owner. Reward teams whose pilots scale into modern operations.

10) *Treating Governance as an Afterthought*

In the rush to deploy AI, governance is often deferred. Guardrails are treated as something to add after systems are live. Without them, outputs drift toward bias, noncompliance, or brand risk. Retrofitting governance creates rework, slows adoption, and increases exposure. When governance is embedded from the start, trust and scale rise together.

Indicators: No defined process to retrain models or detect bias and drift. Privacy permissions and consent flags vary across platforms. Brand and compliance rules are missing from prompts, templates, and workflows. There is no model registry, no versioning, and no owner for approvals.

Solution: Embed governance directly into the path of work. Establish a model registry with a named owner, and training context for key decisions. Capture consent at the moment of collection and block activation if consent is missing or expired. Run pre-release checks for tone, claims, and restricted topics before anything ships. Monitor drift and exceptions with defined thresholds.

Category Three: Talent and Partner Barriers

These problems come from misaligned ownership, skills, and ecosystem choices. Without Marketing Engineers at the core and the right partners tied to the operating model, the system loses momentum before it scales.

11) *Delegating Leadership Away from the CMO*

When AI adoption sits with IT, vendors, or a small operations group, it gets framed as a technology upgrade rather than a redesign of marketing performance. But AI driven marketing reshapes how growth is created. That makes it a CMO mandate, and not a project to delegate. When system design happens outside the CMO's line of sight, marketing becomes a downstream user instead of the AIM OS architect. Tools get installed without a growth agenda, and initiatives drift into technology territory, dramatically weakening the modern marketing impact.

Indicators: System decisions are made without CMO review or sponsorship. Program updates appear in mid-level forums rather than executive settings. Partners describe the work as technology driven instead of performance driven. There is no single owner accountable for the AIM OS or system KPIs.

Solution: Position the CMO as the chief architect of AIM OS. Establish a biweekly AIM OS review chaired by the CMO with key executives, workstream owners, and Marketing Engineers. Lead every executive update with system KPIs alongside outcomes. Require partners to map deliverables explicitly to the four AIM OS zones.

12) *Failing to Build an AI Marketing Culture*

Tools can be installed quickly, but habits change slowly. Without a compelling story for why AI matters in marketing and how it advances careers, teams drift back to familiar processes. AI starts to feel like extra work instead of the new baseline. Pilots succeed technically but stall in mainstream adoption. Incentives and reviews continue to reward traditional outputs, signaling leadership's commitment to business as usual.

Indicators: Teams describe new systems as additional work rather than the way work is done. Pilots succeed but fail to embed in daily workflows. Performance reviews and incentives reward traditional outputs over system outcomes. AI training is optional. There is no shared language for AIM OS flows, roles, or KPIs.

Solution: Make culture the operating platform. Articulate your narrative for why AI matters and how it accelerates careers, then manage to that narrative. Train for real skills such as agent integration, model oversight, data use, and decision design. Rewrite roles so system outcomes are part of core responsibilities. These aren't side projects. Include in reviews what the system learned and how that learning changed the next action. Recognize and reward the behaviors you want, publicly and consistently. When purpose, skills, accountability, and recognition align, AI stops feeling like extra work and becomes the mindset of how work gets done.

13) *Neglecting AI Upskilling*

Platforms continue to add AI features, but without the skills to use them, capability sits idle or gets misapplied. Leaders assume the stack will create performance on its own. Confidence drops, experimentation slows, and vendor help desks replace internal problem solving. Momentum doesn't break because the system fails. It breaks because people tell us they don't feel equipped to run it.

Indicators: AI features exist but rarely appear in live campaigns. Vendor tickets replace in-house troubleshooting. Few teams experiment with advanced capabilities. Training clusters around a handful of enthusiasts. Prompts, templates, and settings are copied blindly, increasing rework.

Solution: Establish a weekly AI Lab focused on live use cases. Keep it hands-on and sixty minutes long. Invite external speakers. Create role-based skill ladders for planners, channel owners, analysts, and creators. Build a shared library of prompts and patterns aligned to brand and compliance. Offer regular office hours with Marketing Engineers so expertise stays inside the team.

14) *Depending on the Wrong Partners*

Many agencies deliver scopes, instead of systems. Without partners who align their work to your operating model, AI adoption fragments into disconnected projects instead of compounding into a performance system structure. Leaders mistake output for architecture. Deliverables ship, but the system doesn't evolve. When partners cannot connect their work to AIM OS flows, they add activity without accelerating capability.

Indicators: Proposals emphasize deliverables over system outcomes. Partners struggle to map their work to AIM OS zones or learning flows. Success metrics track activity rather than sustained performance. Ownership breaks down across decisioning, activation, learning, and governance.

Solution: Shift to a system-first Marketing Engineering partner model. Require every proposal to state its contribution to AIM OS vision. Set shared system KPIs and review them together. Hold a quarterly architecture review with partners. Start every discussion with one question: *What changed in the system that will make next quarter faster?*

15) *Staffing the Shift with the Wrong Talent*

Modern marketing isn't simply "more digital." It is the integration of strategy, data, AI architecture, martech, and operations into a single operating system. Without talent that can bridge these domains, even well-designed structures will struggle. Information doesn't flow. Decisions stall. The system never reaches performance. What's required are Marketing Engineers. It's all about marketing *and* technology expertise that orchestrates marketing specialists and intent, so it moves as one.

Indicators: Hiring prioritizes classic digital roles over hybrid AI system fluency. Teams lack talent that bridges marketing, data, and architecture. Execution slows as design decisions bounce between disconnected specialists. Partners are selected for marketing channel experience or platform knowledge rather than operating structure expertise.

Solution: Appoint a Chief Marketing Engineer (CME) with authority across the four AIM OS zones. Build a high-leverage core of Marketing Engineers who design and run the system so the rest of the organization can move faster without adding headcount. Define three hybrid roles first: Decisioning Architect (signals to next best action), Activation Orchestrator (decision to delivery), and Governance Engineer (speed with safety). Extend this core with a deep bench of trusted Marketing Engineers to provide specialized expertise and scale without locking in fixed cost. As a starting point, target a 1:8 ratio of Marketing Engineers to practitioners in the first year.

Summary

The deeper lesson behind these fifteen barriers is how CMOs evolve to sustain momentum. Each one serves as a signal, showing where legacy reflexes still influence decisions and where the operating model is ready to mature. Recognizing these signals early is what distinguishes teams experimenting with AI from those redesigning how growth operates in the AI era. The shift moves from managing obstacles to designing a system that senses, adjusts, and improves faster than competitors can respond. That is the advantage AIM OS delivers.

Taken together, these patterns of compromises reveal how performance is shaped across the system. Strategic clarity strengthens execution. Data readiness, streamlined platforms, and skilled teams reinforce one another across every layer of marketing. When leaders recognize these signals early and act decisively, detection becomes resilience and momentum accelerates.

Resilient marketing organizations build the capacity to sense friction early and correct it quickly. And they treat detection as a leadership discipline embedded in how the system runs. This enables AIM OS to learn, adapt, and compound advantage over time. The leaders who pull ahead are the ones who design operating models that recover, evolve, and improve faster than the market and the competition.

Leadership Review

1. Which of the fifteen barriers are most likely to appear in our organization?

2. How quickly do we detect when a smart compromise becomes a drag on momentum?

3. Are budget, time, and leadership attention protected for system work?

4. Do partners reinforce the new operating model or pull us back to the old one?

5. If a barrier appears tomorrow, how fast can we detect, correct, and recover?

14

Engineering Marketing's Future

Leading the next era with structure and clarity

This era may be powered by AI, but the pressure marketers feel comes from customers, not technology. Customer behavior is already changing. As intelligence becomes part of how people live, work, and choose, expectations rise across every interaction. And customers don't wait for organizations to catch up. They carry those expectations from one category to the next.

That external reality matters more than any tool decision. Customers are learning what intelligence can do, and once they experience it anywhere, they expect it everywhere. Designing an operating model without accounting for that shift guarantees misalignment, and ultimately a much weaker market position.

Customer behavior follows a predictable progression as intelligence shapes the experience. First it becomes visible. Then it becomes adaptive. Over time, it becomes indispensable. These three horizons describe how expectations evolve and the conditions marketing must be built to operate within.

Together, they explain why customer expectations no longer move in stages. They move in layers. As intelligence reshapes behavior, marketing must operate with the same adaptability. Technology can't be a backdrop. You aren't responding to AI. You are responding to customers responding to AI. It's the environment where growth happens and customers are already changing because of it.

For decades, marketing ran on creativity, intuition, and repetition. That era is fading. The next era depends on intelligence, orchestration, and engineering. And the CMOs who rise will be the ones who understand how consumers behave when deeper intelligence enters the way they explore, compare, and choose. Most teams already feel this shift. And customer actions are changing faster than they can respond. This chapter reminds us that new customer behaviors will define the next decade of marketing leadership.

Customers will expect intelligence in every interaction. They'll expect experiences informed by everything they've already done, to remove the extra work of decision making. They will expect brands to understand context, guide the process, and reduce complexity rather than pushing it back onto them. AIM OS becomes the structure that makes those expectations achievable. This shapes how products evolve, how experiences operate, and how trust is earned across the entire journey.

In the AI era, these shifts will redefine how people explore brands, make decisions, and judge value. Customers will expect clarity instead of confusion, guidance instead of friction, and experiences that recognize intent and adjust in real time. Once they experience that level of intelligence, they won't return to anything that feels slower, less aware, or harder to navigate.

And yet, the fundamentals remain. Relevance still drives choice. Affinity still builds preference. Loyalty is still earned every day. What changes is the framework for influence. Advantage moves from isolated moments to connected flows of data, decisions, and experiences that adapt faster than any team can coordinate on its own.

This is why the book ends where it began, with intelligence back at the center. Not as a tool or technology to acquire, but as a modern capability to architect. Your leadership begins with how you shape the vision, build the teams, and turn intelligence into resilient performance. The CMOs who understand that balance, who bring together precision and creativity, data and design, intelligence and intent, will define the next generation of marketing. They won't wait for clarity. They'll create it. They won't simply manage marketing. They'll engineer it.

AI Is Driving a Wedge Through the Funnel

Artificial intelligence now operates *inside* every layer of the customer journey, reshaping how people learn, evaluate, choose, and commit. The result is a fundamental change in how purchase decisions happens.

Perhaps the best way to illustrate this is with the classic marketing funnel. Moving consumers through a purchase path has long been an intuitive planning framework for marketing teams. Every step on the funnel provides a backdrop for how programs and tactics are designed to guide consumers towards closure. But AI is quickly disrupting these dynamics.

Customers now move continuously between marketing's omnichannel influence and AI mediated guidance. They don't follow a straight path. They explore broadly, narrow quickly, consider options, and reconfirm decisions. They step into AI environments and back out into digital and physical experiences without thinking about the boundary.

Exhibit 25: **The Modern Marketing Funnel**

Consumers Now Navigate Between Omnichannel and AI-Mediated Influence

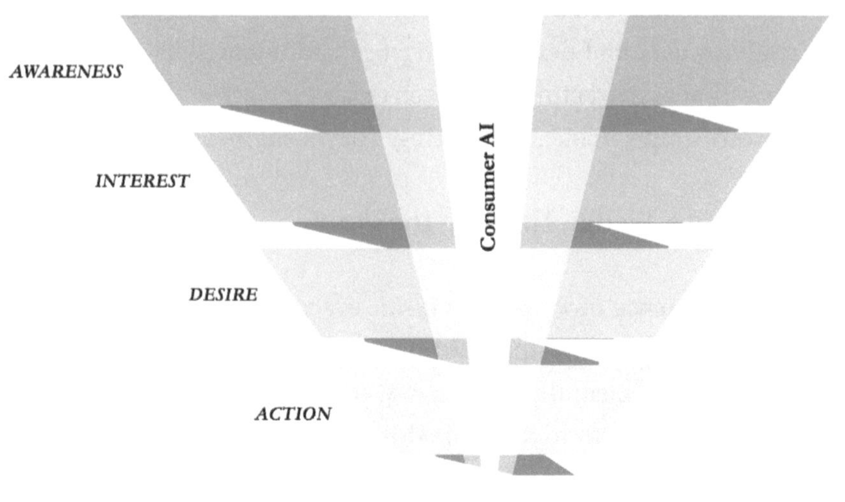

AI doesn't replace omnichannel marketing. It lives alongside it. The challenge for marketing is not about choosing between omnichannel and AI mediated influence. Now it's about learning to operate both simultaneously as one system.

The funnel hasn't flattened or narrowed. But it has become far more dynamic, and far more complex than traditional models were designed to handle. What used to be a well-established planning sequence is now a set of overlapping, intelligence-driven moments. Each layer of the journey can be influenced by AI independently and simultaneously. And customers may move quickly through several decision layers at one time. This is the new normal.

Marketing leaders must now design for a reality where more intelligence lives at every layer of the journey. This is where influence happens inside and outside brand-controlled environments, and where complexity must be handled by systems that learn instead of teams that chase. That is the shift AIM OS is built to support.

The Funnel Still Matters, But It Runs in Two Modes Now

The funnel didn't disappear. It split. Customers still move from discovery to decision to retention, but they do it across two parallel environments. One is the omnichannel world you can manage with paid, owned tactics. The other is the AI mediated world you don't control. This includes assistants, answer engines, and agents that summarize, compare, and recommend on the customer's behalf.

Modern marketing performance comes from executing in both dimensions. Brands still need engaging awareness and great experiences. But they also need representability, meaning the ability to be accurately understood, compared, and recommended by intelligence systems that also increasingly shape the journey. This is the new bimodal reality. If you only operate in one mode, you'll misread where decisions are forming and why outcomes are shifting.

What follows simplifies the journey into four practical bands. Top, mid, bottom, and post funnel. Each still carries familiar objectives. What changes is how AI compresses time, relocates influence, and raises the standard for speed and coherence.

Top of Funnel: Discovery With AI as an Influential Intermediary

At the top of the funnel, customers aren't just searching. They're delegating discovery. AI collapses what used to be repeated exposure, multiple searches, and scattered reading into a small number of synthesized perspectives. The output is more than information, it's *interpretation*. Categories are framed. Options appear early. Decision attributes are elevated or ignored. First impressions now form through the combination of your omnichannel presence and the way intelligence systems summarize the landscape.

That shifts the goal. Awareness is still about reach and authority, but it is also about being legible to systems customers trust to orient them. For example, this is where Answer Engine Optimization starts to matter. This isn't a tactic layered onto existing SEO. It's now a requirement for representability. If your positioning is vague, overloaded with jargon, or inconsistent across assets, you risk being positioned incorrectly or excluded from early summaries. Driving market awareness still matters, but being accurately understood now matters just as much. Top funnel in a bimodal world means you are marketing to people and to the systems speaking to people. Frame this new dynamic as an audience problem. You still market to consumers, but are you treating AI as the new audience it's become, a very powerful new intermediary influencer?

Mid Funnel: Criteria Forms Earlier and Outside Your Visibility
In the middle of the funnel, interest sharpens faster than most teams can manage. AI helps customers translate curiosity into criteria. Models ask follow-up questions. They test assumptions. They refine requirements around personalized needs. AI narrows options, highlights differences, and filters out what doesn't fit specific user context.

Much of it unfolds before a customer visits your site, fills out a form, or enters a nurture path. Intent becomes precise earlier, and it often forms outside traditional measurement. That doesn't eliminate the need for mid-funnel marketing. It raises the bar for what mid-funnel execution must do. They must travel effectively through AI models. And they must demonstrate their offering's context and clarity in ways that survive summarization.

In practice, the mid funnel is no longer shaped only by your manual sequencing. It's shaped by the quality of your *explainability*. Can your differentiation be expressed clearly? Can your claims be validated? Can your strengths stay intact when the customer asks AI to suggest, rank, or find your brand?

Bottom of Funnel: Action Is Pre-Structured Before the Moment of Conversion

At the bottom of the funnel, intelligence shifts from evaluation to execution. Customers ask where to buy, how to buy, what bundle to choose, what terms to negotiate, what constraints to avoid. AI can surface preferred sellers, flag friction, suggest offers, and guide next steps before the customer clicks a button or speaks to sales.

This changes the nature of conversion. The transaction is still yours to win, but more of the decision is structured upstream by AI guided preparation. In many categories, the "moment of conversion" is now the final step in a sequence of reasoning that may have happened across environments you don't control.

That creates a new requirement. Bottom funnel must be designed for speed, clarity, and completion. Pricing logic must be coherent. Offers have to be easy to explain. Paths to purchase have to be frictionless. When customers arrive, the work is already half done. Your system needs to finish it without delay.

Post Funnel: Loyalty Becomes Continuity, Not Just Retention Tactics

Post purchase is where the bimodal shift becomes most visible. In an AI mediated journey, loyalty is less about points and perks and more about continuity. Customers return to brands that remember context, reduce repetition, and advance intelligence forward across interactions. AI can strengthen loyalty by anticipating needs and simplifying repeat decisions. It can also undermine loyalty if AI experiences contradict what the customer has grown to expect from your brand.

This raises the standard for retention. Loyalty is no longer something you activate after conversion. It becomes something the system preserves continuously. Memory, governance, and learning are now part of the relationship. Trust erodes when experiences fail and data becomes unreliable. The question leaders must answer is whether AI is operating as a brand ambassador by reinforcing consistency and values or introducing variability that customers can feel?

What AIM OS Changes Across the Entire Funnel

This is where AIM OS becomes the only plausible response. The challenge isn't that AI exists. The challenge is that the journey now runs in two modes at once. Omnichannel experience still needs orchestration. AI mediated guidance now needs representability and signal integrity.

AIM OS connects the requirements that traditional funnel management kept separate. It links signals to decisions, decisions to activation, activation to learning, and learning to governance. It's how marketing keeps tempo when discovery compresses, criteria forms early, and loyalty depends on consistency.

In a bimodal market, the goal is more than picking the right marketing mix. The goal is to build a system that performs in both environments, with speed you can sustain and trust you can defend.

AIM OS: The Control System for an AI Mediated Funnel

When intelligence operates at every point in the customer journey, the challenge for marketing is continuity. Customers no longer move cleanly from awareness to action. They jump between environments, compress decisions, reopen choices, and arrive with intent already shaped by intelligence systems outside the brand's control. AI accelerates this behavior at every layer, but it doesn't replace the funnel. It changes how the funnel behaves. What was once sequential is now fragmented, faster, and partially invisible.

This is where traditional marketing operations begin to fail. Campaign calendars, channel handoffs, and linear journeys assume order where none exists. They reset context instead of carrying it forward. They respond after behavior has shifted rather than preparing for it. The result is friction customers feel immediately and teams struggle to diagnose.

AIM OS exists to manage this new reality. Not by replacing the funnel, but by operating above it as a control system that preserves coherence as customers move in non-linear ways.

1. Continuity - For When Customers Don't Move Linearly. The defining feature of AI mediated journeys is discontinuity. Customers move between AI tools, digital channels, physical environments, and in person interactions without sequence. Each transition risks losing context. Each reset weakens relevance.

While traditional marketing treats these moments as separate interactions, AIM OS treats them as connected signals. By carrying memory forward, AIM OS preserves intent across moments. It remembers what the customer has already expressed, tested, and ruled out. It allows experiences to pick up where intelligence left off, rather than forcing customers to repeat themselves or restart the journey.

Continuity becomes a structural capability. This makes customers feel recognized. Their decisions feel informed. And operating friction drops without needing to redesign every touchpoint individually.

2. Decision Velocity - For When Intelligence Moves Faster Than Teams. AI doesn't just change where decisions happen, it changes when they happen. Customers arrive with preferences already formed, options already compared, and constraints already defined. By the time they engage a brand directly, much of the decision work is complete. Traditional marketing, built around review cycles and program recaps, cannot keep pace. AIM OS shifts marketing from reaction to readiness.

Instead of waiting for explicit signals, the system prepares actions upstream. Offers, content emphasis, pricing logic, and handoffs are ready before intent becomes visible.

When conditions change, AIM OS adjusts in motion rather than pausing for manual coordination. Velocity becomes an operating characteristic. Marketing acts at the speed of intent, versus the speed of meetings.

3. Control - For When Influence Happens Outside Brand Owned Channels. One of the most destabilizing effects of AI is that influence increasingly happens in environments marketing doesn't control.

Discovery, comparison, and shortlisting occur inside intelligence systems that summarize, evaluate, and recommend on the customer's behalf. This doesn't remove marketing's role. It changes it.

AIM OS allows marketing to shape outcomes without owning every interaction. It aligns how the brand is represented, validated, and reasoned about across external environments. It ensures differentiation travels accurately through systems that customers trust to filter complexity.

Control no longer comes from owning the journey end to end. It comes from governing how intelligence interprets, carries, and applies brand signals wherever decisions are being formed.

What AIM OS Makes Possible

Taken together, these capabilities allow marketing to operate in a way traditional models cannot. Your AI Marketing Operating System is essential now, because it manages continuity across the stages. Here's how:

- Preserves context as customers jump, compress, and re-enter the journey

- Prepares decisions before intent becomes explicit

- Coordinates action without forcing linear paths

- Maintains trust through built-in governance and visibility

- Allows marketing to influence outcomes even when it doesn't control the environment

This distinction matters, because in an AI mediated economy, advantage no longer comes from perfect execution at a single point in the funnel. It now comes from maintaining coherence as customers move unpredictably across it.

This is where the CMO's role fundamentally changes. The job is no longer to optimize channels or orchestrate campaigns in sequence. Your mission is to architect the AIM OS that holds the customer journey together as it fragments and accelerates.

AIM OS gives marketing that system. It turns complexity into flow. It allows creativity, data, and technology to operate in one rhythm. And it enables marketing to move with customers as their behavior evolves rather than chasing it after the fact.

This is why the future of marketing is engineered. It's not because AI is powerful. It's because only engineered systems can sustain relevance when intelligence lives everywhere across the consumer decisions and behaviors.

Marketing Engineers: The Workforce of the AI Era

The last decade introduced marketing technologists. The next will elevate Marketing Engineers. The distinction is essential. Technologists help teams operate tools. Engineers design the systems that turn intelligence into performance. As AIM OS expands across the enterprise, marketing needs people who can work across data, AI, martech, and marketing expertise as one continuous flow. If AIM OS is the architecture of modern growth, Marketing Engineers are the builders who make it attainable.

Today, many of these builders are borrowed. They come from IT, data science, or operations and are adapted into marketing one project at a time. Their patchwork worked during the early stages of digital maturity, but the model has reached its limit. System-level operating models can't run on incidental expertise. The AI era requires talent fluent in marketing experience and technology expertise.

The gap begins upstream. Most marketing graduates still emerge from programs built for a world that moved step-by-step. Research, segmentation, positioning, pricing, campaigns, and analytics remain essential foundations, but they prepare talent for a linear craft, and not for systems that sense, decide, and learn in real time. Marketing has built intelligent systems faster than it has built the workforce capable of running them. This gap is becoming the next source of competitive advantage for organizations willing to close it.

The next generation of CMOs, Strategists, and Marketing Engineers will come from university and college programs that fuse marketing with data science, computer science, and systems thinking. Curricula will expand from marketing principles to modern marketing performance. The degree tracks will be merged disciplines that include:

- *Marketing + Data Science* for behavioral modeling, prediction, and performance optimization

- *Marketing + Computer Science* for API integration, automation, martech, and decision logic.

- *Marketing + Systems Engineering* for architecture, testing, and governance of complex performance systems

These aren't electives. They're the foundations that let emerging talent orchestrate data, engineer systems, and map the decision paths that drive modern brands. Students will read models with the same ease earlier generations defined customer segmentations or construct marketing attribution models. They'll design journeys with system intent. And they'll perform with operating and telemetry precision.

Universities will not reach this future alone. They will need industry leadership. CMOs can accelerate progress by co-designing curricula, funding applied labs, and sponsoring work that requires students to solve real performance problems with modern tools. Graduates from these programs will understand how systems learn, how to measure that learning, and how each interaction shapes the next decision.

But the pipeline will take time, and the need is immediate. The customer horizons are unfolding now. CMOs cannot wait for academia to catch up, nor can they rely on internal patchwork or borrowed capability. Every executive function already depends on specialized external expertise to operate at full capacity. IT leaders don't build everything themselves. Strategy leaders don't develop analytical depth alone. Creative leaders rely on agencies built for craft and expression.

Marketing has reached its equivalent moment. The rise of AIM OS creates a new class of capability that requires a new class of support. CMOs need access to marketing engineers who already know how to translate intent into logic and logic into orchestration. They need teams that can work across data, models, content, and workflow without stitching skills together from across the organization. They need expertise that strengthens the system's learning rate from day one.

Forward-looking organizations are already moving. CMOs are assembling Marketing Engineering capability by blending internal leaders with external depth. Career paths are shifting with them. The most valuable contributors are those who can connect creative intent with algorithmic structure and improve how the system learns, versus just the output of a single campaign.

The companies that build this capability will define the next era of advantage. The choice is no longer abstract. Marketing Engineering must become a cultural and structural commitment.

Technology sets the possibility and talent turns it into performance. As AIM OS integrates across the enterprise, leaders face a key decision on how to build or access the capability, but they can't ignore it. The future belongs to organizations that treat intelligence as a structural discipline and invest in the people who know how to run it.

Marketing Engineering Is the Discipline for the AI Era

Every C-suite leader relies on partners built for their operating reality. Strategy leaders turn to firms designed for clarity and long-range direction. Technology leaders depend on integrators that can deliver platforms at enterprise scale. Creative leaders work with agencies built to express brand and emotion with craft. Each plays a critical role. But none were designed for what marketing has become.

AI marketing systems require a different kind of expertise. They demand fluency across data, models, workflows, orchestration logic, and marketing experience. They operate as connected flows rather than isolated functions. No traditional partner category was built to carry that responsibility. Strategy firms provide insight, but the work now requires active system builders. Creative agencies deliver expression, but the work now requires proven orchestration. Technology integrators deliver enterprise infrastructure, but the work now requires deep marketing context. None were designed to run the intelligence layer of AIM OS.

This is why Marketing Engineering is emerging as a distinct discipline. It fills the structural gap between marketing, technology, and systems. It brings the operating talent and technical depth required to translate intent into logic, logic into orchestration, and orchestration into performance. It's purpose-built to help CMOs design, run, and scale modern marketing systems.

Marketing Engineering doesn't replace existing partners. It completes the ecosystem. It gives the CMO the same kind of specialized support other C-suite leaders already rely on in their domains. CMOs who recognize this shift early gain leverage.

Those who wait will find the gap widening faster than internal teams or traditional partners can close. In the AI era, performance is engineered. Marketing Engineering is the discipline that makes it possible.

The Future of Marketing Is Engineered

This book began with a simple idea that intelligence isn't a tool. Intelligence is a capability. It reveals market patterns we once missed, strengthens go-to-market decisions we once guessed at, and unlocks a wealth of customer signals that once lived in silos. The shift to AI has now reached marketing in full force. Traditional marketing can't keep up with the speed and intelligence of customer decisions or the advantages competitors gain. What began as a function built on layers of marketing eras is becoming a discipline built on flows, models, and systems that learn every day. This is not a trend. It's the new foundation of competitive advantage.

The Modern CMO is the leader who understands this shift and navigates their organization past the hype and into a new era of marketing. AIM OS turns that idea into operating reality.

This is where leadership becomes decisive. The CMO sets the ambition for how the company competes in an AI mediated world. Marketing Engineers design the system that makes that ambition real. No other executive sits closer to the signals that reveal how customers think, decide, and behave. No other function shapes the experience across every point where the brand meets the market. In the years ahead, the CMO's influence will expand because the customer intelligence system it leads becomes central to how the enterprise performs.

The architecture may be technical, but the implication is strategic.

The next era of marketing leadership will be defined by who builds the system that learns the fastest. By who brings marketing and intelligence together. By who designs trust, compliance, and safety into the flow instead of bolting them on later. And by who develops the talent required to run intelligent systems with the same confidence earlier generations brought to media, web, and digital.

The companies that excel understand that intelligence creates value only when it's turned into a working system. Not a collection of tools or series of pilots. A system that decides, adapts, and improves as conditions change. Once that system is in place, it becomes difficult to copy. It shapes how the enterprise allocates capital, how it innovates, and how it earns relevance in a world where every customer carries intelligence of their own.

When the story of this era is written, it won't focus on the teams that experimented with AI at the edges. It will focus on the leaders who engineered intelligence into the core of how their organizations operated. The leaders who treated AIM OS as the architecture of modern growth. The leaders who built the talent, designed the flows, and set the guardrails that allowed the system to improve day after day.

Marketing has always been about relevance. In the years ahead, relevance will be earned through intelligence at a pace instinct alone cannot match. This is the moment to build the system that will shape the next decade. This is the moment for marketing to take its place at the center of enterprise performance.

Because the future of marketing won't be powered by AI.

It will be engineered by you.

Conclusion

As this book ends, your work begins. *The Modern CMO* gives you a new way to see *how* marketing operates in the AI era. The shift is already underway and the leaders who rise now will be the ones who understand that marketing is stepping into a new operating reality.

Global investment and adoption of artificial intelligence will continue to accelerate. Reasoning, decisioning, and generative capabilities perfectly mirror the blend of intuition and analysis that has always defined marketing. Yet most organizations remain built on layers of legacy capabilities stacked across decades. This is a structure that can no longer keep pace with this new AI era. Marketing now requires an operating model designed for intelligence rather than inherited workflows.

Boards and CEOs feel this urgency as much as CMOs do. Their expectations for AI often begin with efficiency and productivity, but the real shift will reach much further. This turn is about competitive edge. We already know that AI is reshaping the environment your customers live in. People already turn to AI platforms to filter information, compare choices, and navigate decisions in real time. Buying behaviors are changing and their expectations now come from every intelligent experience they touch. When an AI agent becomes the front door to your category, your funnel is no longer a funnel. It's becoming a negotiation with an intelligence you don't control. Some insist we are moving to a machine-to-machine marketplace (M2M). While the future continues to unfold, what is certain is that marketing leaders insisting on protecting yesterday's marketing approach will watch relevance erode quickly.

That is why the pressure feels different. While the work of marketing has evolved continuously, the structure has remained largely the same. This is why the gap between them continues to widen. We like to say this is the moment to get your shift together. That is your mission now. The power of generative intelligence and the span of AI agents is transformative. You already know this is an inflection point.

What remains unknown is how quickly organizations will act. There will always be reasons to wait. These include new AI announcements, changing market trends, and shifting business priorities. Don't get distracted. The fundamentals of marketing are still essential, and this means staying relevant with customers comes first. Your modern marketing advantage will come from the performance of your AI Marketing Operating System (AIM OS) and how quickly it learns customer behaviors, how well it adapts to them, and how reliably it improves with each cycle. Every cycle of learning sharpens your OS. But with every month of delay, you lose compounding ground to competitors who continue to learn and improve their system.

The encouraging part is that once you see the shift clearly, you can lead it. And that brings us to the most important question. How does the CMO evolve to meet this era with confidence?

Seeing the shift is only the beginning. The real work now is what you choose to build. Modern CMOs are being pulled into a different kind of leadership that brings vision together with technology, execution together with engineering, and ambition supported by operating discipline. You will need to navigate the challenges with a steadier hand, build the structures your teams can rely on, and shape an AI operating fabric that elevates every part of the function. Above all, you will face the leadership choice that defines your moment to design the future of marketing.

Your leadership starts with how you view your role. The modern CMO is no longer defined by the outputs marketing produces. Now you are defined by the operating model that makes those outputs possible. You don't need to master every AI technology or understand every agent under the hood. Your work is to become the architect of how modern marketing operates.

This kind of architecture isn't foreign to you. It draws on the instincts that have guided your best work for years. These instincts are purpose, clarity, alignment, collaboration, and insight. When you design clean workflows, when decision logic becomes explicit, and when data connects rather than fragments, you give your team what they have been missing. A modern structure they can trust and a system that allows them to run at the speed they are capable of. You are literally leading them into the future of marketing, that is already here.

The reality is that you're more prepared for this moment than you might think. You have been designing marketing systems your entire career. But the stakes are higher now and the scope is larger, yet the same leadership principles apply. This is the next level of the CMO role, where your architecture becomes the force that turns intelligence into highly competitive performance.

The modern CMOs who are winning today aren't pushing harder or adding more activity. It means seeing yourself as the conductor of your AIM OS that stretches across product, programs, channels, research, and brand. When you lead marketing this way, something powerful happens. The rhythm changes. Decision speed accelerates. Handoffs shrink. Engagement rises. And teams stop battling the org charts and start trusting the rhythm of an aligned mission.

Your leadership gives team members the sense that the system is working with them rather than against them. Said another way, it replaces complexity with confidence. It removes the friction that exhausts teams and gives them an environment where they can collectively do their best marketing work.

Modern CMOs also need a new kind of technical partnership. You personally don't have to stitch the martech stack or calibrate the AI agents yourself. This is the work of Marketing Engineers. Marketing Engineering is the discipline that turns your strategy into an AI operating reality by connecting technology and data, so everything finally moves as one.

What makes this especially encouraging is that some of the people already on your team in Ops, Martech, Data, Product, and IT are part of the way there. You aren't starting from a blank slate. You're elevating capabilities that already exist, giving them the structure and purpose they need.

When Marketing Engineering becomes embedded in the way marketing runs, your vision stops living on slides and starts appearing in the market. Your team gains the marketing technology partners who enable their best work. Because now the AI foundation beneath them is finally strong enough to support their ambition.

The future of your function will be shaped through the four performance capabilities you now lead. You will guide how signals become decisions through Adaptive Decisioning. You will unlock scale through Autonomous Activation. You will champion continuity through Experience Loop Intelligence. And you will protect the integrity of the system through Intelligence Governance. Together, they give marketing a new performance rhythm, and a new way to win.

Every leader understands the weight of 'this is the way we've always done it'. The gravitational pull of business as usual is very strong. Remember, you are now simultaneously leading in two worlds, the old traditional model and the new, modern model. But you now have a map for recognizing that pull early and for leading your teams beyond it. Breaking legacy habits doesn't require confrontation. It requires clarity and steady guidance.

When you clean up one workflow, or resolve data issues, your team will feel the difference immediately. You will guide them forward with conviction and confidence, one structural improvement at a time. And those improvements will compound and become the proof that the new model works.

Modern marketing leadership also depends on a shared operating logic between the CMO and the CEO. When you share a single operating view of how marketing works, everything becomes easier. Decisions move faster. Investments stay consistent. And accountability now means the organization aligns around one rhythm instead of competing agendas.

Many CMOs underestimate how much CEOs want this clarity. They're looking for the model, the structure, and the language that allow them to support intelligent marketing with confidence. When the two of you move in sync, the organization feels the alignment. The work gains speed. And the market notices.

The Modern CMO is defined by how marketing works.

You're designing marketing as a performance model built for the pace of modern customers rather than a collection of activities. In an environment driven by intelligence, structure becomes the force multiplier. Structure amplifies talent, and a strong operating model amplifies your leadership. This is what modern leadership looks like, and it's fully within reach.

You don't need to overhaul your entire function in a quarter. You only need to get started. The most effective CMOs begin with a single marketing capability, they design, build, and run it through an AIM OS model. Then they iterate, improve it, and move to the next objective, and then the next. In the AI era, progress doesn't come from large scale transformations. Instead, it comes from small structural wins that build confidence. When confidence spreads, it lifts teams. It lifts morale, and it changes how people see what's possible.

One journey, one workflow, and one action at a time is how modern CMOs engineer belief. Not just belief in the system, but belief in the work and belief in themselves. Once the wins begin to stack, the organization feels the shift. The model begins to hold. And the future becomes something you can build toward rather than something you react to.

What you're building is something genuinely special. You're engineering how intelligence moves through your organization at scale and speed. You're designing how your company will compete in the years ahead. You can do this because you already have the marketing experience, the customer focus, and now the operating model to lead with clarity.

Modern marketing is a discipline where creativity, data, and advanced technology move together in real time. Marketing Engineers bring this to life every day, fusing strategy and creativity with intelligent systems. This creates new levels of customer segmentation, journeys, and personalization. The diversity of this work is extraordinary, and it moves faster than traditional team structures could ever execute. It blends invention with orchestration, experimentation with leadership, and it invites teams to achieve what once felt out of reach and then extends their reach even further.

Inside every modern marketing organization that embraces this shift, you will see people working together to solve complex AI enabled challenges. You will see collaboration that pushes technology farther than any individual could achieve. You will see brand builders, data thinkers, and engineers shaping the next generation of marketing faster than anyone expected. Innovation has always driven brand relevance and loyalty, and today it fuels competitive advantage.

This is why the role of the CMO matters so deeply. AI isn't the lead story for marketing. The Modern CMO is the story. You inspire new forms of talent. You give Marketing Engineers a place to grow their craft. You create the environment where intelligence thrives, creativity scales, and customer experience becomes dynamic. You begin with traditional structures and a bold AIM OS ambition. You are now shaping the progression that will define your organization's future.

The business world is watching. Customers, CEOs, boards, and teams are looking to see how marketing will lean into this next era. The CMOs who embrace this moment will be seen as enterprise builders who lead their companies to market in new and evolving ways through systems that learn, adapt, and improve every day.

The Modern CMO is defined by *how* marketing works, and you are engineering the roadmap for a new generation of performance.

Index

C

D

F

G

H

Author

Robert Painter is the Founder and CEO of Martechture, a prominent consulting and marketing engineering firm that helps CMOs design operating models built for the AI era. A former CMO of IBM Global Business Services and Senior Vice President of Corporate Marketing at Cognizant, he has led global organizations through brand strategy, operational transformation, and multi-billion-dollar enterprise growth.

Painter has spent his career at the intersection of marketing, technology, and enterprise services, guiding teams through successive waves of transformation across the discipline. His work with Fortune 500 companies and high-growth innovators informs the core of this book.

Through Martechture, he and his team of Marketing Engineers help organizations build and operate their AI Marketing Operating System, strengthening data foundations and embedding intelligence into how marketing decisions are made and scaled. He brings a practitioner's perspective shaped by decades in the field and a conviction that the next era of marketing advantage will be determined by how well marketing is engineered to perform.

www.ingramcontent.com/pod-product-compliance
Lightning Source LLC
Chambersburg PA
CBHW020721150726
48196CB00036B/888/J